INTRODUCTION TO ENERGY TECHNOLOGY

INTRODUCTION TO ENERGY TECHNOLOGY

Marion L. Shepard
Associate Professor

Jack B. Chaddock
Professor and Chairman

Franklin H. Cocks
Associate Professor

Charles M. Harman
Professor and Associate Dean
of Graduate School

Department of Mechanical Engineering and Materials Science
School of Engineering
Duke University
Durham, North Carolina

ANN ARBOR SCIENCE
PUBLISHERS INC
P.O. BOX 1425 • ANN ARBOR, MICH. 48106

Second Printing, 1977

Copyright © 1976 by Ann Arbor Science Publishers, Inc.
P.O. Box 1425, Ann Arbor, Michigan 48106

Library of Congress Catalog Card Number 75-36284
ISBN 0-250-40123-1

Preface

The age-old necessities of life are food, clothing, and shelter. The 20th century has dramatized a fourth—energy. Energy starvation of the technological complex that maintains modern society may be soon as crucial a problem as feeding the world's hungry. Indeed, energy starvation could well precipitate more widespread food starvation.

Solutions for the energy crisis are strongly dependent on the technology of how energy is used. To make a physical change in the world it is necessary to use four resources: energy, matter, space, and time. How well a task has been performed can be measured in terms of the amount of fuel consumed, the mass of material used, the space occupied, the hours of labor to accomplish it, and the ingenuity with which these resources are utilized. Squandering of irreplaceable energy resources, waste of materials, or large expenditures of space and time cannot be tolerated if the necessities of life (and some of its very desirable luxuries) are to be provided for all.

Technology addresses itself to the efficient utilization of these four ingredients of physical change. We have witnessed an era of cheap energy, and, as a result, have developed an industrial complex that is highly energy-intensive. The era of cheap energy is now ending. It is being replaced by one in which the populace will necessarily become energy conservation conscious; first because of the rising cost for energy, but later because of the dire consequences in placing additional stresses on our biosphere, already showing serious signs of the strain.

Our way of life has been growth—exponential growth. Now we face limits to this growth; exponential growth cannot go on forever. If technology has brought forth this confrontation between further growth and those problems that make such growth a questionable benefit, should we look to technology for the solution to that confrontation? We think the answer is an unqualified yes, yet recognizing that the solutions will contain many nontechnological aspects. We find that many others who have recently given this question serious, informed thought, agree.

Dr. Philip Handler, president of the National Academy of Sciences in an address at Duke University's 50th anniversary celebration, spoke apprehensively of the serious problem of growth. "I have difficulty in facing the future with equanimity, difficulty in imagining a happy outcome," he said. Nevertheless, he expressed an optimistic viewpoint—that of the power of technology to further improve man's use of the earth's resources.

The principal conclusion of the three-year, $4 million Energy Policy Project of the Ford Foundation, as we see it, is the necessity of slowing our energy growth. A proposed reduced growth curve is called the *Technical Fix* scenario. "This scenario reflects a conscious national effort to use energy more efficiently through engineering know-how, that is, by putting to use the practical, economical, energy saving technology that is either available now or soon will be." That is how the Ford Foundation Report, A TIME TO CHOOSE, describes its proposal.

To slow energy growth, and practice energy conservation at its best does not mean we can end our search for new energy supplies. Fossil fuel resources, particularly oil and natural gas, will be seriously depleted by the year 2000, even with an adopted national goal of reduced energy usage. The potential for, and technical problems associated with, the development of renewable energy sources, including solar, wind, tidal, ocean thermal currents, and geothermal are being seriously discussed and investigated. Nuclear energy from fissionable isotopes, and perhaps, in the future, by fusion of the hydrogen isotopes deuterium or tritium, must be looked to as a part of the world energy supply. Some replacement for oil as the predominant source of transportation energy must be found.

For some years now, it has been our purpose to make available to Duke University students, from all disciplines, an introductory course for the serious study of the technology of energy supply in the world of today and tomorrow. We have chosen to call that course *An Introduction to Energy Technology*. This book is a part of the outcome from our efforts.

There is a great reluctance by students in humanities and social sciences, and some by students in life sciences, to elect the study of technical subjects. We have tried to present the material here in such a way as to overcome that reluctance, and to provide the students with a knowledge of energy sources, uses, and methods of conversion sufficient for an understanding of our energy crisis and an appreciation of its technical problems. We have also attempted to use the course as a vehicle for stimulating beginning engineering students to consider a career development in energy problem solving. Should others find the book useful in these objectives, or others, we would be pleased to learn of their experiences with it and suggestions to improve it. In this regard, we are well aware that throughout the text both English and metric units are intermingled. We have chosen to do this because it reflects the real engineering world that presently exists in the U.S. Thus, to force uniformity of units in an area where such uniformity does not yet exist would be artificial and could serve to confuse the student rather than to enlighten him.

The authors gratefully acknowledge the major contribution of Linda Hayes in preparation of the original manuscript and Pauline Clow who assisted her.

Marion L. Shepard
Jack B. Chaddock
Franklin H. Cocks
Charles M. Harman

Table of Contents

Chapter 1

U.S. Energy Usage—
Past, Present, and Future

THE ENERGY CRISIS

The "energy crisis" is usually expressed in terms of two major concurrent symptoms brought on by an exponential growth rate in energy usage. The first is the generation of massive amounts of pollutants that affect mankind and the total biosphere in unknown and possibly injurious ways. In terms of today's knowledge, it is impossible to predict over the long term of several hundred years, either the effect on global climate, or the effect on human health. As in the case of cigarette smoking, however, there is enough evidence at this time to make possible a sound judgment that the injurious effects are a major problem, if not a crisis. The ever-increasing usage of energy poses an inevitable energy-environment conflict, regardless of the source of energy, be it fossil fuel, nuclear, or even solar. With improved technology these environmental effects can be remedied, to a certain extent, but at a price. The costs can only be estimated because knowledge of pollution effects is incomplete. That the price is high can be recognized from estimates provided by the Environmental Protection Agency to clean up the air pollution in the United States. For the year 1970 alone, these estimates ranged from $6–8 billion.

The second symptom of the energy crisis is the depletion of our most desirable and easily obtainable fossil fuels, particularly oil and natural gas. In 1971 this country used 15.2 million barrels of oil per day, or 5.5 billion barrels for the year. In the same year we used 22.7 trillion cubic feet of gas. These two fuel sources provided 77% of our national energy supply. Of the remaining 23%, better than 18% was from coal, 4% from hydropower, and less than 1% from other sources, including nuclear. Overall our energy usage in 1971 was the equivalent of 34.5 million barrels of oil/day.

1

Our nation's energy budget has been recently growing at an annual rate of 4.5%. This is the equivalent of 1.5 million barrels of oil per day. In the past few years, about one-third of that growth has been accounted for by natural gas, nuclear power, and coal. The remainder of 1 million barrels per day each year was oil, *imported oil*, mostly from the Middle East. At the current price of about $10 per barrel, this means an additional import cost of 10 million dollars per day or $3.65 billion per year. In 1973 our oil imports amounted to 6.2 million barrels per day. In terms of today's oil price, this would cost our nation about $22½ billion.

Based on the present estimates of our recoverable domestic reserves of oil and gas (known and undiscovered) and our growth rate of usage in the past few years, we will deplete those reserves by the years 2000–2050. Figure 1, reproduced from a report[1] by the Joint Committee on Atomic Energy in 1973 should enable the reader to grasp an understanding of this nation's energy dilemma. To the left of the line for the year '73 is the supply and demand for domestic fossil fuels showing, mostly, a surplus. To the right of the line is the situation we face for the future in terms of the depletion of our domestic resources, the reliance on imports, and the necessity to develop, through technology, large amounts of energy from nonfossil energy sources. The '73 line may be considered the division between the era, now ending, of relatively cheap and available domestic fossil fuels, and the emerging era of a requirement for an ever-increasing supply of imported fuels and new energy technologies.

MEASURING ENERGY USAGE

Many different units are used to measure energy. Some of these may be familiar to you from every-day use, but even then, it may be difficult to relate energy in one set of units to another.

Let us define two energy units commonly used in science and engineering.

1. A British thermal unit (Btu) is very nearly the energy required to raise the temperature of one pound of water one degree Fahrenheit (°F).
2. A Joule is exactly equal to the energy when one watt of electricity is used for one second.

The watt is a unit of power, which is defined as the time rate at which energy is used.

$$\text{Power} = \text{Energy/Time}$$
$$1 \text{ watt} = 1 \text{ Joule/sec}$$

The common unit of electrical energy, and the one by which your electric utility bill is determined is the kilowatt-hour (kwh). The

B/D OIL EQUIVALENT vs YEARS

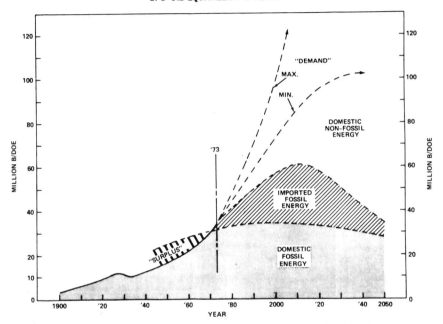

Figure 1. U.S. energy supply and demand from 1900–2050.

large size of today's electric generating stations is such that their power ratings are often expressed in Megawatts (Mw).

$$1 \text{ Mw} = 1000 \text{ kw} = 1,000,000 \text{ watts}$$

Even the Mw is not sufficiently large to conveniently express the total energy usage of the United States. A common choice for this role is the "Quad," which stands for a quadrillion Btu, or 10^{15} Btu. Table 1 presents some useful conversion factors between the Quad, and other methods of expressing energy usage.

Table 1. Useful Conversions for Expressing Energy Usage

1 bbl (42 gal) of oil = 5.8 million Btu
1 ft³ of natural gas = 1035 Btu
1 ton of coal = 26 million Btu
1 kwh of electricity = 3412 Btu
1 Quad/yr = ½ million bbl oil/day
1 Quad = 1 trillion ft³ gas
1 Quad = 40 million tons of coal
1 Quad = 300 billion kwh

HISTORICAL GROWTH IN ENERGY USAGE

The first energy shortage must have occurred during Paleolithic times when man's hunting techniques were so well developed that large animals eventually disappeared. Out of this shortage, no doubt, developed methods of agriculture that precipitated the Neolithic revolution. Thousands of years later, in the fourth century A.D., the Roman empire encountered an energy crisis brought on by a man-power shortage. Up to that time, human muscle had powered the ancient world. This crisis was solved by the harnessing of water power, which came into use throughout the empire. A great flour factory mounted on a steep grade of the Roman aqueduct ground 28 tons of flour in 10 hours—enough to feed 80,000 people.

Twelve hundred years later, a wood shortage occurred in England. Wood was then used as both construction material and a fuel. "The preparation of a ton of bar iron from ore required twelve loads of charcoal, or about eight beech trees one foot square at the base."[2] The government imposed conservation measures. In one act of 1593, beer exporters were compelled to either return the original barrels or to come back with foreign clapboard suitable for making the same number of barrels as had been exported. The result of this crisis was the substitution of coal for wood as fuel. Coal was cheap and plentiful. Technical problems prevented immediate conversion, but as these were solved new manufacturing techniques were invented which laid the foundation for the Industrial Revolution.

The need for more and more coal led to the development of steam engines to pump the water from mine shafts. The need to move tons of coal from mines to factories brought on a revolution in transportation. First this led to horse-drawn railways and, by the beginning of the nineteenth century, to the first steam locomotives.

Now we are facing an energy crisis in fossil fuels. Perhaps this too will result in another technological revolution in new sources of power. It seems certain that there will be, as well, a new social order.

In Figure 2, the annual energy consumption of the United States is shown for the years 1850–1970. In this period the total energy usage increased from about 2.5–68 Quads. Some interesting history and drama of the development of this country lies in this figure. One example is the stock market crash of 1929 and the ensuing depression of the 1930s.

In 1850, about 90% of the energy was provided by wood, while in 1970 wood supplied less than 1% (but still more than nuclear energy). Each new fuel enters the graph at a steep slope, not displacing the older fuels then in use, but adding to them. Note that since 1935 the fastest growing fuel usage has been in natural gas. Nuclear power just entered the energy picture in 1970, but came in at a very rapid rate.

The growth rate in usage of fossil fuels is seen more clearly in Figure 3 for the years 1920 to the present. Coal has had a curious growth pattern. Its consumption was 13.6 Quads in both 1912 and 1970. It has had highs of 17 Quads during World Wars I and II and lows of 9–10 Quads in 1932 and 1958. Recently the trend has been slightly upward.

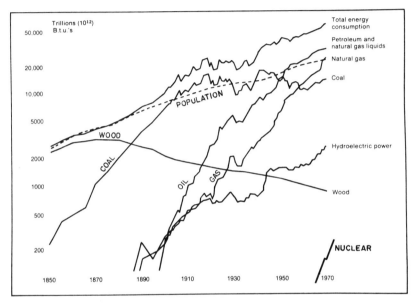

Figure 2. An historical record of energy supply and consumption in the United States, 1850–1970.[3]

It was stated earlier that in 1971 oil and gas provided over three-quarters of our energy supply. This is displayed graphically on the right-hand side of Figure 3.

ENERGY USE AND THE ECONOMY

The relentless march of energy usage in the USA has outpaced the rate of population growth. This can be seen in Figure 2. In the 20-year period from 1950–1970, population increased from roughly 150–200 million or about 1.5% per year. In the same period, total energy consumption increased from 34–68 Quads, a doubling in 20 years which corresponds to a 3.5% growth rate per year. In recent years, 1965–1973, this difference is even more pronounced, with the energy growth rate increasing to 4.5% a year, while population growth slowed. The outcome of such growth rates is more energy available per person, which, not surprisingly, results in a higher standard of

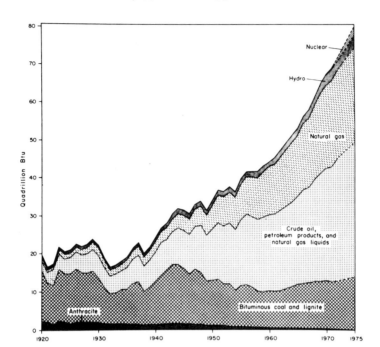

Figure 3. Trends in sources of energy supply for the United States, 1920–1975.[4]

living and a greater gross national product per capita. The historical growth rate in GNP/capita is given in Table 2.

Table 2. Energy/Capita and GNP/Capita Growth Rates.

Year	1880	1910	1940	1970
Energy/capita[a]	100	180	190	330
GNP/capita ($)	800	1300	1700	3600
GNP/10^6 Btu ($)	8	7	9	11

[a]in millions of Btu.

A value of 330 million Btu/capita is the approximate equivalent of the caloric intake of 80 humans. One way of visualizing this per capita energy usage would be as the equivalent of 80 slaves working for every man, woman, and child in the USA. Most of the increase in GNP per capita has resulted from higher energy use per person, but, as shown by the bottom line of Table 2, there has also been some increase in efficiency.

The relationship between energy usage and productivity is universal. Figure 4 shows the relationship for 46 nations around the world. The United States uses more energy than any other nation. An often repeated statement is, "The United States with 6% of the world's population, accounts for about 35% of the world's energy usage." What is not stated is that with that energy we produce more than 35% of the world's goods and services. The efficiency of national conversion of energy to GNP can be roughly compared in Figure 4; those countries lying above the solid line have the more efficient economy. Note that Switzerland uses less than one-third the energy per capita of this nation but has a GNP/capita that is 60% of ours. This may be partly due to the excellent technological development of cheap hydropower.

While in recent years energy usage has been growing in the U.S. at a more rapid rate, it has grown even faster in Western Europe and the developing nations of the world. It is obvious from Figure 4 that the standard of living in many Southeast Asian, South American, and African countries are at very low levels. The demand for raw energy supplies has become highly competitive and strained. The world demand for oil imports has grown at the rate of 10.8%/yr over the past decade. This requires a doubling of output by oil exporting nations in a little less than 7 years. Since 1970 the United States' increase in energy consumption has been almost totally dependent on oil imports. Oil furnished 46% of our total energy consumption in 1973, and in that year 35% of the oil was imported.

In view of the above facts, it should not be surprising that the Organization of Petroleum Exporting Countries (OPEC) began to negotiate higher prices. In the four years ending January 1, 1974, the price of oil rose from about $1.30/bbl to $8/bbl, an increase of 515%. It has since undergone price rises to more than $11/bbl. "As a consequence, the total oil revenues of the OPEC members soared from 7.8 billion in 1970 to $23 billion in 1973. They could reach $90 billion in 1974. To put these numbers in perspective, the total revenues earned by all developing countries from exports in 1972, including oil, amounted to about $19.6 billion." The Ford Foundation Energy Policy Project report,[6] "A Time to Choose," from which the above quote is taken, has among its conclusions these statements:

> "The dramatic price increases in the world oil market over the past four years are more than a temporary aberration on the supply and demand charts. According to our judgment, they represent a fundamental shift in the power relationships between the world's industrial powers and the oil exporting nations.—Rising oil prices have seriously aggravated the already existing world economic woes—inflation and currency instability."

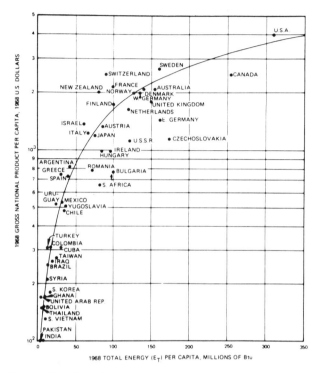

Figure 4. Gross national product per capita vs. total energy (E_T) per capita consumption, 1968.[5]

TYPES OF ENERGY USAGE

The most common broad breakdown of energy usage is into the four categories of Electrical Energy Generation, Residential and Commercial, Industrial, and Transportation. In Figure 5, the energy consumption in these four categories is shown for the years 1950 and 1970. The vertical scales in units of millions of barrels of oil per day are the same for both years, which gives a further graphical illustration of the total energy growth rate for this 20-year period.

Electrical energy generation differs from the other three categories in that it is not an end use in itself. As can be seen from Figure 5, the preponderance of electricity goes to Industrial and Residential and Commercial usage. Note that the greatest increase has been in electrical generation. While the overall energy use increased at a rate of 3.5% a year for the period 1950–70, electrical energy usage increased at double this rate of 7% a year. We have been moving toward an electrical energy economy.

You should also notice that in the generation and transmission of

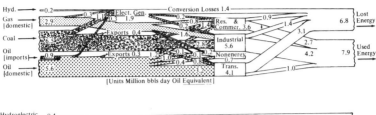

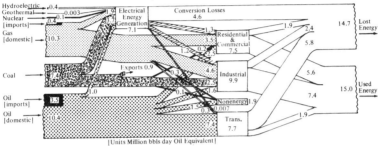

Figure 5. U.S. energy flow diagrams[1] for 1950 (above) and 1970 (below). At the
left side are labeled sources of supply, in the center the end use is indicated, and at the right is the efficiency of usage.

electricity, about 70% of the energy content of the original fuel is lost.
In other words, electricity is 30% efficient. Before 1968, the improvements in technology of electrical energy generation were able to
more than offset inflation and fuel price increases, so that electric
utility rates declined. In 1926 the average cost to the ultimate consumer was nearly 3¢/kwh. By 1940 this had declined to 2¢/kwh, and
in 1967 was about 1.3¢/kwh. Since that time, due to fuel cost increases and the requirements for reduced pollution, the rate has
about doubled to most electricity consumers.

The other low-efficiency energy user is transportation with about
75% of the fuel energy (almost totally oil) being lost. The internal
combustion engine, which consumed about 85% of the transportation
energy, was developed more for low power-to-weight ratio rather
than for efficiency. Anyone driving a late-model automobile knows
further what antipollution devices have done to fuel economy.

Residential and Commercial energy usage accounts for 28% of the
total raw energy usage, including that which comes through electrical generation. The distribution of the residential energy usage
in the household is illustrated by Figure 6. It is significant to note
that space heating requires 65%, space heating and water heating
78%, and space heating and cooling plus water heating 83.5% of the
total. Almost all of the heating is done by gas and oil or their derivatives. Conservation efforts, which reduce space heating requirements

through lowering thermostat settings, better insulation and storm windows, could achieve significant savings in fossil fuel usage. Clearly "turning off the lights" is not what reduces your household utility bill. In a house heated by oil or gas, the lighting is still a small percentage of the electricity usage. Table 3, taken from Healy[7] gives the percentage distribution of electrical usage for a representative American home using 10,000 kwh/yr.

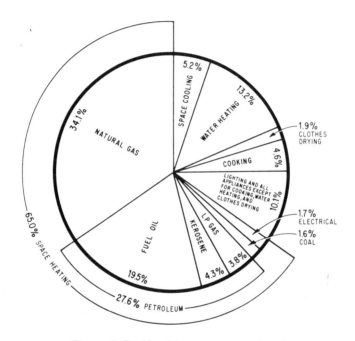

Figure 6. Residential energy usage, 1970.[5]

Table 3. Residential Use of Electrical Energy.[7]

Use	kwh	%
Cleanliness (including hot water heating)	5200	52
Food storage and preparation	2800	28
Lighting	1000	10
Service appliances; iron, vacuum, etc.	500	5
Entertainment; television, radio, etc.	500	5
Total	10,000	100

Relating energy use to people's lives is a complex task of tracing the flow of energy through society. The Ford Foundation Energy

Policy Project has developed a rough estimate of certain categories of indirect household energy use. Table 4, reproduced from that report[6] illustrates the indirect uses.

Table 4. Indirect Energy Use per Household (10^6 Btu).[6]

Income Group[a]	Food	Autos	Housing	Appliances	Govt. Services	Other	Total
Poor	38	35	10	6	65	199	353
Lower middle	65	82	11	7	65	199	549
Upper middle	79	121	13	9	65	544	843
Well-off	94	147	16	10	65	763	1095

[a]The income groups are defined in terms of average annual income as follows: poor—$2500; lower middle—$8000; upper middle—$14,000; well-off—$24,500.

The households in each income group are; poor — 11.8 million, lower middle — 27.6 million; upper middle — 12.6 million; and well-off 13.4 million. Table 4 shows clearly that as household income increases so does indirect energy consumption. The direct energy usage figures are similar being approximately 210, 290, 405, and 480 million Btu/yr for the four groups in order of increasing income.[6]

Industrial energy usage is the largest among the three end consumption types in Figure 5, with nearly 37% of the total. In fact, this is also an indirect usage, since industry produces goods and services which ultimately go to the American consumer or for export. Energy is an input needed to produce these goods and services. For the most part industrial energy is used "in or by machines, equipment, buildings, and appliances, which in general constitute the capital stock of the economy. Energy cannot be eaten or worn. It is used through the medium of capital goods, from complicated industrial machinery to the lowly household iron. The rate of growth in this capital stock, its changing mix, and changes in efficiency with which it transforms energy into goods and services are fundamental determinants of the growth in U.S. energy demand."[6]

In general, our industries have been energy-intensive. The higher costs of fuels now provide economic incentive for technical improvements in efficiency of conversion of energy to consumer products and services.

SCENARIOS FOR THE FUTURE

Projections of energy usage for the future are a vital and necessary part of planning both for the private and public sector of our society. Electric utility industries are now experiencing nearly a 10-year span between decision to build a large nuclear power generation station and its being brought "on-the-line" for service to customers. President Ford and his advisers must look even further into the future in developing a sound national energy policy. The Ford Foundation Energy Policy Project was conducted over a period of three years at a cost of $4 million. Its final report is called "A Time to Choose — America's Energy Future" and is recommended reading for anyone seriously interested in understanding the U.S. energy problem and rational policies for solving it.

In making an analysis of energy choices, the Ford Energy Policy Project[6] constructed three different versions of possible energy futures for the U.S. through the year 2000. These are illustrated in Figure 7, taken from the final report of the project. The three scenarios are called "Historical Growth," "Technical Fix," and "Zero Energy Growth."

The Historical Growth Scenario is simply a continued projection of the past trend in energy usage. Energy usage grows at a 3.4% rate annually, the average rate for the period 1950–1970. The situation in

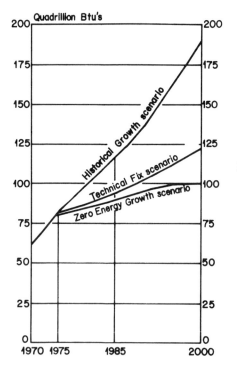

Figure 7. Ford Foundation Energy Project—scenarios for the future.[6]

1990 would be about as shown in the energy flow diagram of Figure 8, when we would be using 66.8 million bbl/day oil equivalent or 142 Quads of energy. Figure 8 is taken from the same JCAE study[1] as Figure 5, and again uses the same vertical scale. A comparison with the 1950 and 1970 energy flow diagrams in that figure illustrates the continued exponential growth rate. Such a growth assumes a continued vigorous national effort to enlarge energy supply, and little to no change in our energy appetite.

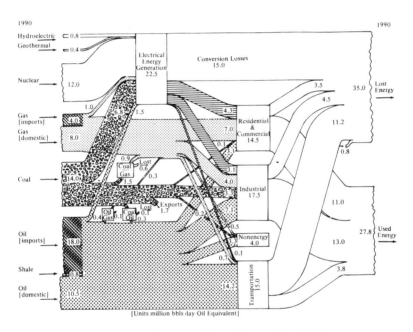

Figure 8. U.S. energy flow diagram for 1990.[1]

On the source side in Figure 8, fossil fuels would still provide the large bulk of energy supply, although intensive nuclear development has been projected for 12 million bbl/day oil equivalent, or 18% of the total. This scenario also calls for 18 million bbl/day of oil imports and 4 million bbl/day oil equivalent of gas imports: a price we may not be able to afford. Some other technological developments appear in Figure 8 including coal gasification, coal liquifaction, geothermal energy, and some shale oil production. Efficiencies of conversion have not improved much in this all-out effort to keep up with demand.

The Technical Fix Scenario of Figure 7 shows a marked reduction in projected energy growth. The annual energy consumption increases at an average rate of 1.9%/yr. It is proposed, however, that it can provide essentially the same level of energy services (miles of travel, quality of housing, manufacturing output, etc.) as in the Historical Growth Scenario. This is to be accomplished by practical, economical, energy-saving technology that is available now or will soon be developed. This conservation-oriented energy policy should provide benefits in every major area of concern; reduction of costly oil and gas imports, avoiding shortages, protecting the environment, and allowing time for the technological developments necessary to increase energy supply without disastrous consequences.

The Zero Energy Growth Scenario projects a slightly lower growth rate than Technical Fix to about 1990, whereafter energy supply remains constant at 100 Quads. This future would call for some sociological changes, away from energy-intensive industries with emphasis on making things, to a more service-oriented society. Even with this energy scenario there is a large energy appetite to be satisfied, one-third larger than at present. It is the belief of the Advisory Board for the Energy Policy Project that it is both technically and economically feasible to achieve stability in energy consumption while continuing healthy economic growth. They note that "environmental concerns or resource constraints may force us to such a policy, whether we like it or not."

A much more modest effort at energy projection was made by a small faculty group participating in the NASA/ASEE Systems Design Summer Faculty Program. Their study,[8] completed in 11 weeks at Auburn University, was concerned with applications of solar energy to the energy crisis. As a required step in their analysis, they also prepared three projected energy scenarios for the years 1971 to 2020.

Figure 9 presents the "Upper Bound" and "Lower Bound" scenarios. The Upper Bound consumption curve is roughly the same as the Historical Growth Scenario, calling for 206 Quads of energy by 2000 and 377 in the year 2020. Note the large demands for new energy supply from solar, nuclear, synthetic gas and oil, and geothermal. In spite of these sources the requirements for imported oil and gas are staggering. The study group concluded that this scenario would have the following impacts:[8] a) an energy crisis (and related crises) would truly occur; b) all nonenergy R&D efforts would have to be curtailed to concentrate on a crash R&D program on energy; c) the nation would be on the verge of international bankruptcy by 2020; d) over $1 trillion in capital would be required to construct the nuclear plants projected prior to 2020; and e) pollution standards would have to be relaxed and nuclear waste storage would become an increasingly touchy problem.

The Lower Bound Scenario in Figure 9 is nearly the same as the Zero Energy Growth Scenario in Figure 7. The total energy consumption is 98.3 Quads in the year 2020. Note the reduction in dependence

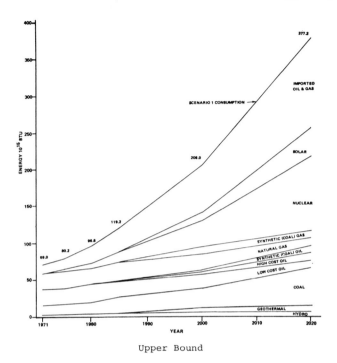

Upper Bound

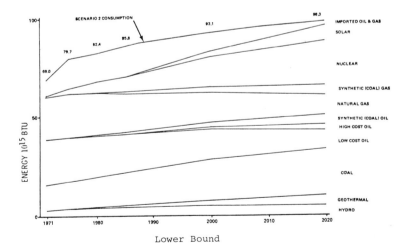

Lower Bound

Figure 9. TERRASTAR Energy Scenarios 1 and 2.[8] The Upper Bound Scenario projects a continued exponential growth rate. The Lower Bound Scenario projects a decreasing growth rate, approaching a zero rate by 2020.

on foreign oil to near zero by that year. The probable impacts of this scenario are stated as:[8] a) a population control program would be required, as well as stringent immigration policies; b) strong federal control of energy consumption will be necessary—probably fuel rationing; c) industrial growth and GNP could be seriously affected, and our standard of living will be somewhat retarded; d) nuclear power will be installed slowly enough to permit sufficient time for waste storage problems to be solved; and e) pollution levels can be easily controlled.

As might be expected, the study recommends a more rational midcourse of planned growth. This is shown in Figure 10. It was the judgment of the group that such a growth rate in energy supply, "will have a stabilizing effect on society and the environment." The probable impacts of Scenario 3 were given as:[8] a) the increase in energy growth though slowed, would permit high-energy users to maintain their usage, whereas low-energy users could benefit from a higher level of per capita energy consumption; b) the rate of expansion necessary for all energy sources can be managed without any crash programs; c) oil and gas imports would grow worse until 1985 but decline thereafter; and d) at this moderate rate of expansion in energy usage, pollution abatement devices can be perfected and economized without excessive cost.

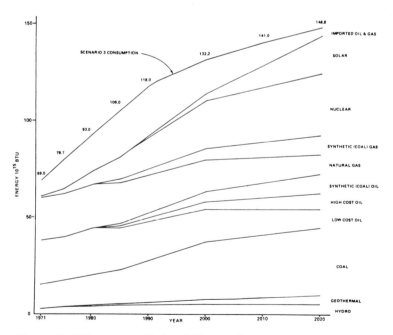

Figure 10. Midcourse or Rational Energy Scenario of TERRASTAR.[8]

McGeorge Bundy, President of the Ford Foundation had this to say in the Foreword to the Energy Policy Project Report.[6]

> "The report itself owes much to the advice and criticism offered along the way, and the remaining differences, though sometimes sharp, seem in their own way to underline the central message of the study — that it is truly *Time To Choose*. The measure of the agreement is accurately registered in the general statement of the Advisory Board on Major Issues. There *is* an energy crisis. It did *not* come and go in 1973-74. It *will* last a long time. We *do* need an integrated national policy."

REFERENCES

1. "Understanding the National Energy Dilemma," Joint Committee on Atomic Energy, JACE Print, (Washington, D.C.: U.S. Government Printing Office, 1973).
2. "Man's Age-old Struggle for Energy," *Natural History Mag.* (October, 1973).
3. Hottel, H. C. and J. B. Howard. "An Agenda for Energy," *Technol. Rev.* (1972).
4. Risser, H. E. "The U.S. Energy Dilemma: The Gap Between Today's Requirements and Tomorrow's Potential," *Illinois State Geological Survey*, No. 64 (1973).
5. "The U.S. Energy Problem," *Inter Technology Corp.*, Vol. 1 Summary and Vol. 2 Appendices — Part A, NTIS PB 207 517 (November, 1971), and PB 207 518 (November, 1972).
6. *A Time to Choose — America's Energy Future*, Energy Policy Project of the Ford Foundation, Final Report, (Cambridge, Massachusetts: Ballinger Publishing Co., 1974).
7. Healy, T. J. *Energy Electric Power and Man*, (San Francisco, California: Boyd and Fraser Publishing Co., 1974).
8. *Terrastar*, Final Report, NASA CR 129012, (Auburn, Alabama: Auburn University, School of Engineering, 1973).

Chapter 2

Energy Supply

INTRODUCTION

Our civilization had its roots in the Asia Minor-Persian Gulf area. Today, strong waves again emanate from this region as the Arabs impose embargoes and otherwise regulate the flow of oil that powers western civilization. The previously predicted large supplies of fuels suddenly changed to conditions of apparent shortage. The resulting shock caused many charges and countercharges to be made. It is this complex and many-sided subject of energy supply that we wish to examine here.

A whole array of energy sources can be considered. Historically, wood has been a principal source of energy. Its use and importance peaked perhaps a century ago as it was displaced by the fossil fuels. Today, energy supply and fossil fuel resources are almost synonymous in the minds of many. Certainly other energy sources are important today, and surely, as we shall see, some will take on much greater importance in the future. Nevertheless, it is the fossil fuel that dominates today and it is this classification that we shall examine first.

Two basic uncertainties are involved in estimating geological energy resources. The obvious one is involved with the fact that these mineral deposits lie hidden beneath the earth's surface. Making an accurate appraisal of these deposits to determine their extent and quality is exceedingly difficult. Problems generic to this facet will be discussed in terms of the minerals considered. The other aspect of resource estimation concerns the problems associated with assessing whether a particular mineral deposit is of a sufficiently high grade and sufficiently accessible to be counted as a resource. Changing economic considerations and technological capabilities can individually or in concert alter the assessment of whether or to what extent a deposit is a resource. The combination of these factors render resource estimates which differ astonishingly by a factor of 2 or, in many important cases, much more.

Before proceeding further, however, some attention should be paid to the definition of terms used in the classification of resource estimates. This will be done with the caution to the reader that uniformity of these definitions is by no means standard.

Reserves will be considered first. Reserves refers to deposits that are recoverable under current economic and technological conditions. Reserves may be be divided into two major classifications—those which have been located geographically and those which lie still undiscovered. Deposits in known fields or locations may be further classified into proved, probable, and possible reserves. The quantities in a particular deposit move from the least certain to the most certain (but not absolutely certain) classification of proved reserves as more is learned about the deposit. Proved reserves may only represent a fraction on the order of 10% of the total resource. This may be true, for example, even in a reasonably well-developed oil formation.

Resources are those deposits which can be made available under certain arbitrarily assumed technological and economic conditions. Frequently, a resource is defined to include those deposits that can be recovered or can be expected to be recovered at a cost of within 1.5 times current energy costs using forseeable technology. Submarginal resources, which can only be expected to be recovered at costs exceeding 1.5 times current costs, are also included in some resource estimating. It can easily be seen that resources may be suddenly propelled into the reserve category by nothing more than a step increase in unit price of the deposit.

The relationship between reserve and resource in terms of certainty and economic feasibility is illustrated in Figure 11.

PETROLEUM

Petroleum is by far the most critical of the fossil fuels in today's world because of the severe demands for it and because most of the nations that are the large petroleum users are not able to meet their needs with their own supplies. Since petroleum is probably uppermost in everyone's mind when energy supply is mentioned, it will be examined first.

An estimate of the geographical distribution of the world energy supply of oil is illustrated in Figure 12. This graphically indicates the unenviable position of the United States with respect to the future, especially when compared with some other areas of the world. It should be noted that in terms of *proved* reserves, the Middle East has 53% of the world oil and the United States, including Alaska, only 5%. Also, it is interesting to note that during the past 25 years, the proved recoverable reserves in the United States have

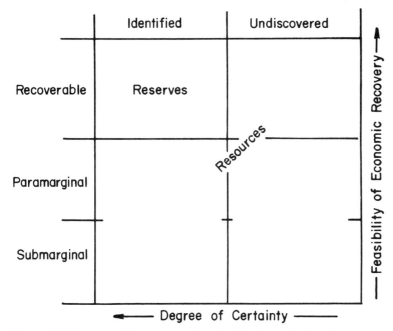

Figure 11. Conceptual diagram for the classification of mineral resources.[1]

doubled but its percentage of the world's proved recoverable reserves has plummeted from 31%.

It has been estimated that the original amount of oil in the United States, including offshore areas, was 2800 x 10⁹ bbl. (A barrel, the standard for petroleum measurement, is 42 gal., not the 55 gal contained in a standard steel drum). A number of estimates of the recoverable oil have been made. Within the past 20 years, estimates have ranged from about 150 x 10⁹ bbl to the top estimate of 590 x 10⁹ bbl. Most estimates place this figure at between 200 and 400 x 10⁹ bbl. To date, only about 90 x 10⁹ bbl have been used. Under these circumstances, one might conjecture that all is well and that, at most, oil supply is to be another generation's problem. A careful look at Figure 13 will dispel this optimism, however. If the lower estimates prove correct, U.S. oil production is at or about at peak and we soon face a decline in production. The recent uniform oil production in the United States is a somewhat ominous indicator that the lower curves may be more accurate than the upper one. Another presentation of total U.S. oil production based on 1971 statistics is shown in Figure 14. The ultimate production of 170 x 10⁹ bbl was arrived at by using a "discoveries per foot of well drilled" method and other production, discovery, and proved reserves data. One is led to

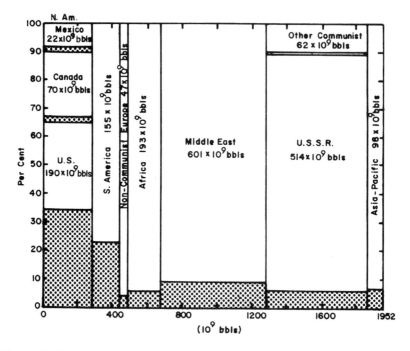

Figure 12. Graphic representation of Jodry estimate of the world's ultimately recoverable crude oil. The shaded areas at the foot of each column or sector represent quantities consumed already.[1]

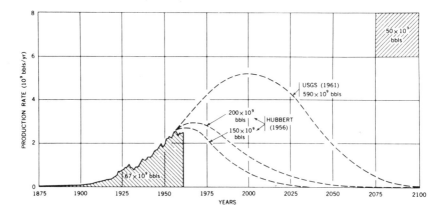

Figure 13. Comparison of complete cycles of U.S. crude oil production based upon estimates of 150–200 and 590 billion barrels.[1]

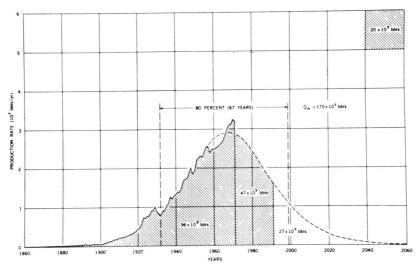

Figure 14. Complete cycle of crude oil production in conterminous United States as of 1971.[1]

surmise from this figure that the oil shortage is not just the result of a conspiracy among oil companies to raise prices.

The increased difficulty, subsequent to the early 1950s, of finding oil is illustrated in Figure 15. It should be noted that the average depth of finding oil is increasing and that well costs are by no means proportional to depth, but increase rapidly per foot at greater depths. A 30,000-foot well is only 6 times deeper than a 5,000-foot well, but the cost is not 6 times as much, nor 12 times, nor even 36 times, but more than 120 times. In the past, an average of 160 bbl of oil were discovered per foot of well depth. A future average rate of 120 bbl/ft is projected. Drilling costs then will also be increased by this reduced productivity as well as by the previously mentioned increased average well depth.

A very considerable difference exists between the amount of oil originally in place in the ground and the amount which is projected to be recovered or recoverable. Oil in the ground is contained in porous rock. The amount depends on the porosity of the reservoir rock and the ease with which it flows to the well depends upon how well these tiny pore spaces are connected. Another factor influencing the flow of oil in the reservoir rock is the oil viscosity. These widely variable factors are chiefly responsible for the great variation in recovery from existing fields. Recovery of up to 50% of the oil originally in place has been obtained in some east Texas fields while only 10% has been recovered in some Oklahoma fields. An overall view of the problem of oil recovery from fields in the United States is given in Table 5. Estimates such as this do not include the effect of a technological breakthrough, but anticipate improvement in current methods and technology.

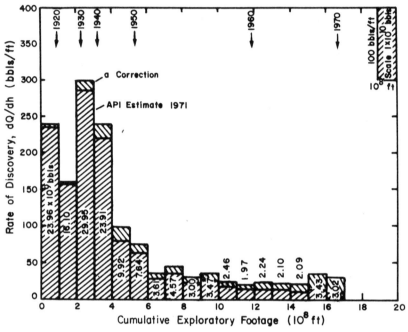

Figure 15. Average discoveries of crude oil per foot for each 10^8 feet of exploratory drilling in the conterminous United States from 1860–1971.

Table 5. Estimated Recoverability of Oil from Fields in the United States.

Produced	25%
Recoverable under present conditions	10%
Recoverable with present technology, greater cost	12%
Recoverable with improved technology	13%
Unrecoverable	40%
Total oil originally in place	100%

Oil is recovered from the oil field with varying degrees of effort. Primary oil recovery refers to that oil which runs to the well from the surrounding media due to pressure and gravity effects. On the average, around 30% of the total amount of oil in place will be recovered by this method.

Additional oil may be produced by using fairly simple techniques. This is known as secondary recovery. The most widely used method for obtaining secondary recovery is to flood the oil field with water. The presence of the water tends to float an additional amount of oil out of the media. This is usually not a very costly procedure and results in an average additional yield of around 10%. Another level of effort, known as tertiary recovery, may be instituted when secondary recovery methods fail or become ineffective, or production may be terminated if analysis suggests that prospecting for new oil may be more economically effective.

Tertiary recovery methods include anything done to increase yield after water flooding. Many methods or concepts for tertiary recovery exist and various of these methods or combination of methods are under intensive development. Current tertiary recovery technology may be capable of increasing yields by 12%, and methods yet to be developed may double this yield. Most research progress in tertiary recovery techniques is held confidential because the companies involved wish to maximize the benefits to themselves of improved technology. A nontechnological aspect of tertiary recovery is that generally all nearby wells in a particular field benefit when tertiary recovery methods are used. This usually means that, when these wells are under different ownership, all owners must agree on cost and benefit sharing.

Some of the more promising methods of tertiary recovery are as follows:

1. **Hydrocarbon miscible.** In this method, a light hydrocarbon is injected into the well, usually liquified petroleum gas. This dissolves into the remaining oil and forms a less viscous fluid which is more readily displaced from the pores of the oil-bearing media. Water injection is also used in this method. Difficulties include distribution of the solvent in the well and the problem of loss of valuable solvent in the process.

2. **Carbon dioxide miscible.** Carbon dioxide is injected rather than the LPG as in the above method. It is miscible with some crudes and acts in manner similar to the hydrocarbon miscible method. It has the advantage that the carbon dioxide gas may be a very inexpensive by-product of another process.

3. **Water miscible.** Water-base solvents are employed. A surfactant solution is used which will form a microemulsion with oil, and a drive fluid of water made viscous with polymer addition completes the process. Loss of a portion of these solutions in the process is the drawback.

4. **Thermal methods.** These methods use heat to thin the oil and make it flow more easily to the production wells. Steam is used

most extensively. Injection of oxygen and underground burning have been considered. The Russians have exploded nuclear bombs in oil wells and effectively increased the yield.

The costs and effectiveness of these methods is shown in tabular form in Table 6. Clearly, petroleum produced by tertiary recovery methods will be more costly.

Table 6. Effectiveness and Cost of Secondary and Tertiary Oil Recovery Methods.[2]

	Normal range of recovery improvement-percent original oil in place	Cost increase $/bbl of added oil
Secondary recovery		
Waterflood	10–50	.35– .50
Steam	10–60	.75–1.25
Tertiary		
Miscible methods	30–80	.75–1.25
Thermal	40–70	1.25–1.50

NATURAL GAS

Natural gas resource projections are especially difficult since it occurs under two different sets of conditions. The gas called "non-associated gas" is natural gas which occurs without contact with crude oil. "Associated gas" occurs in contact with crude oil either as a "gas cap" overlying the crude oil in the dome of the reservoir or as a gas dissolved in the crude oil at the extremely high pressures existent in the reservoir. Fortunately, about 80% of the natural gas in place is generally deemed recoverable. Extraction is stopped when pressures reach uneconomic levels. Also, a portion of the gas in place exists in reservoirs too small for economic exploitation.

Resource estimates for the United States for natural gas vary considerably due to assumptions concerning oil production, gas to oil percentages, and the relative likelihood of discovery of future nonassociated wells. Using a figure of 1050×10^{12} ft^3 as the total recoverable amount of natural gas originally in place, a production cycle can be shown as in Figure 16. It can be seen that a total of 414×10^{12} ft^3 have been used. A very high percentage of the gas produced prior to 1945 was actually an undesired by-product of oil production. As such it was burned or "flared" at the wellhead. This waste of this important resource was stopped in Texas by law in the late 1940s. Even today though, sattelite pictures taken at night over the region of the Middle East to central Eurasia show the location of each of the oil fields because of the tremendous illumination from flaring natural gas.

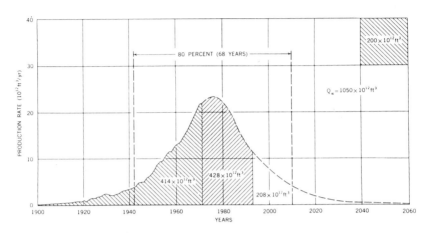

Figure 16. Estimate as of 1972 of complete cycle of natural gas production in conterminous United States.[1]

Current reserves total about 400 x 10^{12} ft^3 and the predicted remainder is relatively small. The large pipeline system for collecting and marketing the gas make it presently a high-demand commodity. Another factor related to its demand is government regulation of price at the wellhead. The Federal Power Commission set the extremely low value of 16¢/1000 ft^3 originally in 1954. (The standard cubic foot is taken at atmospheric pressure and 60°F). Since that time, the price has been allowed to rise to only 20¢/1000 ft^3. Gas not subject to this regulation (not in interstate commerce) now costs about 2½–5 times this much and liquified natural gas imports cost about 5 times this amount. Obviously there is little incentive to stimulate production of regulated gas and shortages are to be expected until owners can receive a competitive market price. Also, quite obviously, increased natural gas costs are certain in the future.

A table of the comparative energy from alternate fuels is shown below.

Fuel	Unit	Energy per unit (10^6 Btu)
Natural gas	1000 ft^3	1.035
Crude oil	42 gal barrel	5.8
Bituminous coal	ton (2000 lb)	26.2

Natural Gas Liquids

Natural Gas Liquids (NGL) are the hydrocarbon components of natural gas that become liquid at surface conditions of about 20°C and 1 atmosphere pressure. They are generally reported independently of gas and crude oil statistics but form a significant category. In 1971 the production of NGL was 19% of the crude oil production. The estimated cycle of NGL is shown in Figure 17 for comparison to crude oil and natural gas production cycles.

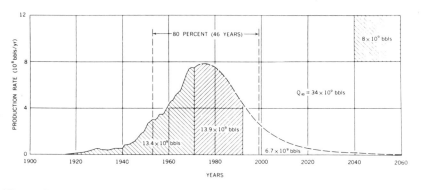

Figure 17. Estimate, as of 1972, of complete cycle of natural gas liquids production in the conterminous United States.[1]

COAL

Coal is by far the most abundant fossil fuel. Although deposits are virtually worldwide, coal is concentrated in the earth's upper northern latitudes. Coal is principally carbon, but it contains varying amounts of volatile material, ash and moisture, which causes the heating value to vary from around 14,000 Btu/lb for good grade bituminous to 7000 Btu/lb for some lignites. The estimated distribution of coal is shown in Figure 18. As noted, this includes coal in veins down to only a 12-inch thickness and at depths down to 6000 ft. It includes lignite and subbituminous coal existing in beds of at least 30-inch thickness. Further it includes only 50% of the coal actually estimated to be in place as this is all that can normally be removed. The rest is needed for mine support columns.

These estimates may be accurate, but there is considerable question as to whether all this coal should be considered a resource. A 1-foot thick seam at a 6000-foot depth would certainly be most unattractive. In the past, using manpower and burrows, seams of at least 3 ft in thickness were required. Modern mining equipment can utilize seams with thickness down to 14 inches. If one considered anthracite and bituminous coal in seams at least 28 inches in thickness and subbituminous coal and lignite in seams at least 5 ft thick, the resources would be decreased to only 26% of the values shown in Figure 18.

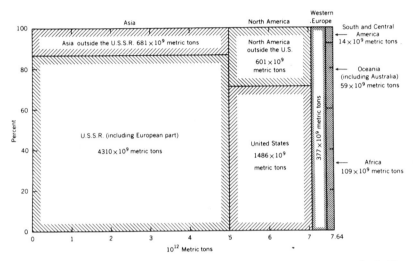

Figure 18. Estimates of initial world resources of recoverable coal in beds 12 or more inches thick occurring at depths of 6000 feet or less.[1]

Estimated complete cycles of U.S. coal production have been projected and are shown in Figure 19 for the two cases discussed. It might also be remarked that the world cycle is fairly close in time to this projected cycle for the U.S. Certainly we have a vast resource of coal which has the capacity of contributing to man's energy needs for generations. Coal provides 97.5% of the nation's fossil fuel energy reserve. The proved reserves of coal are perhaps only 20 x 10[9] tons, a rather modest part of the total resource.

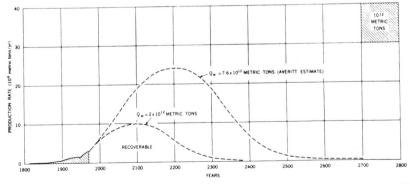

Figure 19. Two complete cycles of world coal production based upon Averitt higher and lower estimates of initial resources of recoverable coal.[1]

SHALE OIL

Shale oil is another potential energy resource which has not been commercially tapped as yet. World deposits of oil shale are shown below as

	Oil in Place in Shales	Proved Reserves
North America	2,250,000	80,000
Canada	50,000	
U.S.	2,200,000	
South America	800,000	50,000
Europe	78,000	30,000
Africa	100,000	10,000
Asia	110,000	20,000

where only shale containing at least 10 gal of oil per ton and in mineable thicknesses is considered as "in place." Proved reserves considers only accessible portions of major deposits of high-grade shale which is available using current technology. All are given in millions of barrels of oil. Two observations are quite evident: a) the oil shale deposits are very significant—world reserves are approximately half of the world reserves of oil, b) perhaps more significant, more than two-thirds of these deposits are in the United States.

The Green River Formation, concentrated in about a 16,000 square mile area in Colorado, Utah and Wyoming, contains the bulk of the world oil shale. The formation contains up to $2,000 \times 10^6$ bbl of oil. High-quality deposits (yielding over 25 gal/ton and running at least 10 ft thick) contain an estimated 590×10^9 bbl. A high proportion of this can be utilized economically today using proved mining techniques. In addition, 1150×10^9 bbl are contained in lower-grade deposits yielding at least 15–25 gal/ton. At current international oil prices, these deposits are emerging also into contention as an economic resource of energy. Problems preventing intense utilization of these resources are technological because of limited water availability in the region for processing and political because of environmental protectionists not wanting to see this relative wilderness area become populated and developed.

TAR SANDS

Tar sands are another natural resource of energy. Worldwide, they form a reserve about one-third the size of the oil shale reserve. They are far less widely distributed than are other fossil deposits. Most are in Africa. A large deposit exists in Canada, which is under commercial exploitation. The only other consequential deposits are in the United States. These deposits contain about one-fourth the oil of the Canadian deposits and represent only about 5% of world deposits. It might be remarked that they are probably only about 2% of the size of the shale oil deposits in the United States.

FISSION FUELS

Fission fuels generally are in abundant supply overall. Uranium which occurs in nature as uranium oxide (U_3O_8), is the principal source of fissionable material for nuclear power production. The free world resources of uranium are shown below:

Non-Communist World Resources of Uranium in 1970
(tons x 10^3 of U_3O_8 at a cost of $10/lb or less)

	Reasonably Assured	Estimated Additional	Total Resource
United States[3]	390	680	1070
Canada	232	230	462
Other	458	345	803
Total	1080	1255	2335

As can be seen, the United States has a generous proportion of the world uranium resource. The Atomic Energy Commission (AEC) has made a careful inventory of these deposits in the United States. The major differences in tables such as the one above are in the determination of how much is available at a specified cost. The deposits are all far from pure and the cost is largely associated with refining. Ores under consideration in the above table may contain as little as 100 ppm of U_3O_8.

Although the uranium resource above is reported in terms of U_3O_8, actually the material used in light water reactors is U^{235}, an isotope of uranium, which makes up only one part in 140 of the natural substance. Through the use of plutonium, a synthetic element produced in the fission reaction, approximately 1.5% of the potential energy of uranium can be realized rather than the approximately .7% available directly from the ore. With this 1.5% energy utilization, uranium oxide produces 4.5 x 10^{11} Btu/ton, making it an extremely highly concentrated source of energy. When breeder reactors become an operational reality the energy utilization may be expected to be on the order of 80% or 50 times greater, thus expanding by this factor the energy potential of uranium. Another way of looking at this would be to say that the effective fuel cost would be reduced by a factor of 50, or conversely, that fuel costing 50 times as much would be equally economic. Under these circumstances uranium ore at $500/lb would become equivalent to today's $10/lb ore. Using ore of up to this cost would provide an assured supply of 80 x 10^9 tons, an amount sufficient to provide current United States electrical usage for 2500 years. An additional amount 3 times this great probably exists. In addition, seawater, which contains .003 ppm of U_3O_8 may be a source of uranium at a cost of perhaps only $30/lb. The resource of uranium is truly staggering. Thorium, an element similar to uranium, and abundant, is also fissionable and is a potential extender of the fission fuel resource.

A last remark concerning proved reserves of uranium should be made. In 1970 the proved reserves were around 52 x 10^3 tons, an amount providing only a 10–12-year supply. It is unlikely that proved reserves for longer periods of future supply will accrue because of the cost of proving the reserves. This does not indicate a potential danger in the supply, but is only a matter of economics.

NUCLEAR FUSION

Nuclear fusion using deuterium and/or tritium, hydrogen isotopes, provides the potential for a virtually limitless source of energy. Two of the most promising of the hypothesized fusion reactions are the deuterium-tritium and the deuterium-deuterium reactions. Helium, nuclear particles, and a large amount of energy are released in these reactions. Most experts in this field, however, do not expect that controlled nuclear fusion can be accomplished on a commercially significant scale during this century. The energy per unit weight using these "heavy" hydrogen isotopes is about 4 times that from the fission of U^{235} and about 10 million that from the combustion of gasoline. Each cubic meter of sea water has the energy equivalent of 1360 bbl of oil due to its deuterium content; so if the fusion process can be brought under efficient control, energy supply would seem to be permanently assured.

SOLAR ENERGY

Solar energy in its various forms is a continuous source of energy supply. The earth intercepts about 10^{14} kw of solar power. About 10^{13} kw is received at the rotating surface. The resultant energy causes plants to grow, the winds, the rains and the temperature variations. Some of the usable forms of resulting energy, expressed in terms of power available on a continuous basis, are shown below:[4]

Source	Power (kw)
Photosynthesis	10^{10}
Available ocean heat	10^{10}
Available wind energy	10^9
Hydropower	10^8

where the entire earth is considered and annual averages are used in reporting the power available. An obvious direct use may be made of the solar radiation for heating and cooling buildings and other domestic purposes, even though the flux is low and variable on a daily and seasonal basis. An additional comment should be made concerning hydropower, since this form is and has been of significance in meeting energy needs.

HYDROPOWER

Hydropower provides a small but nevertheless significant portion of the energy resource of the United States. It may come from either a run-of-river site or from a reservoir produced by a dam. The power produced at any site is proportional to the rate of flow of water times the vertical distance through which it falls. Power produced from a reservoir is far more useful than that produced from run-of-river sources since the reservoir provides energy storage, making it possible to meet varying demands regardless of the flow of the stream at the moment. It has been estimated that sites technically capable of being developed, economic feasibility not withstanding, could produce

48.5×10^6 kw—available 90% of the time
$64. \times 10^6$ kw—available 50% of the time
$86. \times 10^6$ kw—based on mean flow

Developed and installed capacity or capacity under construction is approaching the portion available 90% of the time. The total capacity for full development is 122.1×10^6 kw, which will produce 513×10^9 kwh. The indicated installed capacity exceeds the average capacity because installations to utilize average flow are often not feasible. It might be mentioned that hydropower development in the U.S. is quite variable among the various regions and river basis. The Tennessee river provides 92% of its potential hydropower while the Potomac provides only 2%. The majority of the hydropower comes from the Pacific coast region although the eastern region of the country has more fully developed its hydropower potential. As hydropower development proceeds, developers must look to progressively less favorable development sites. Some sources predict little future development of hydropower in the United States.

WIND ENERGY

Wind energy can be produced from the world's winds which have been estimated to have an average total wind power of 10^{17} kw. The world average wind speed in the lower atmosphere is about 9 m/sec. This is equivalent to approximately one-half kilowatt per square meter of "windmill" area. The efficiencies of extracting energy from wind will be discussed in Chapter 10. Certain areas are more well suited to wind power. Both high winds and constant winds are desirable.

OCEAN HEAT

Ocean heat is available to produce power due to the temperature difference between the surface and the deep ocean water. Warm surface water overlies cold water at depths of as little as 300 ft near the equator. A heat engine can be made to operate utilizing the temperature difference, and although the thermal efficiency (percentage of heat exchanged that is converted to work) would be quite low, perhaps 3%, the large volumes of water available would make possible the power cited earlier. Areas such as the Gulf Stream would be particularly favorable for producing this form of energy.

PHOTOSYNTHESIS

Photosynthesis, the process of producing plant material utilizing energy from the sun, may also be harnessed for producing energy for man. Today, about 6% of the earth's land area is under cultivation. Only about 1% of this agricultural photosynthesis is actually used by man. Much of the remaining 99% could be used as fuel or to produce fuel. The unharvested vegetable matter and garbage could be anaerobically digested to produce methane, as well as fertilizer. It has been estimated that approximately 10% of current world usage of natural gas could be produced this way. Phytoplankton production in the oceans produces perhaps 100 times the amount of photosynthesis material produced in agricultural production and, with effective harvesting techniques, could some day become a significant source of energy. Of course other sources exist such as the forests which, until relatively recent times, produced in wood the primary energy source for man. Bagasse, the residue stalk from cane sugar production is one of the few examples of an agricultural product systematically used as an energy source. Use of agricultural products to produce energy is extremely unattractive from a cost point of view. Figure 20, shown below, illustrates this point dramatically.

TIDAL POWER

Tidal power has the potential to become an important source of energy in selected locations. Such plants have operated over the centuries. Early records indicate working tidal mills in the eleventh century in Europe. They have been in use in this country, and an ill-fated federal effort to develop significant tidal power was made in

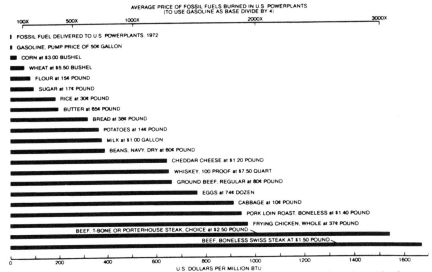

AVERAGE PRICE OF FOSSIL FUELS BURNED IN U S POWERPLANTS
(TO USE GASOLINE AS BASE DIVIDE BY 4)

Figure 20. Comparison of prices of energy derived from fossil fuels with those derived from feed and food.[5]

this country in the 1930s at Passamaquoddy Bay in Maine. A large modern installation exists in France and an operational system is operated by the Russians today. The world's total tidal power has been estimated at 13,000,000 kw. Not associated with the tides, but brought to mind when considering the seas and coastlines, is the energy of waves. It has been estimated that wave energy of significant magnitude could be harnessed at specific locations through the use of focusing horns and other devices.

GEOTHERMAL ENERGY

Geothermal energy is available from the earth's hot interior. On the average, only .063 watts/m^2 of energy is conducted to the surface. However, geothermal areas, areas where this flux is much higher, could provide 60 million kw of electric power for decades. Operational plants are in existence in the United States (California) and other sites are being actively investigated. A well-developed site in Italy has attracted much attention.

SUMMARY

Energy is abundantly available from many sources, but dependence on petroleum must yield in the near future to other forms, perhaps lacking in the convenience and economy to which we have become accustomed.

REFERENCES

1. "U.S. Energy Resources, A Review as of 1972, Part 1," Committee on Interior and Insular Affairs U.S. Senate (Washington, D.C.: U.S. Government Printing Office, 1974).
2. "Conservation and the Efficient Use of Energy, Part 3," Joint Hearings, Committee on Government Operations and Science and Astronautics, House of Representatives, (Washington, D.C.: U.S. Government Printing Office, 1973).
3. "Energy Research Needs," Part 1, Resources for the Future, Inc., prepared for NSF (October, 1971), pp. 11-73.
4. *Oceanus,* Vol. XVII, (Woods Hole, Massachusetts: Woods Hole Oceanographic Institute, 1974).
5. *Exxon USA,* Second Quarter, (Houston, Texas: Exxon Corporation, 1974).

PROBLEMS

Always state assumptions clearly and indicate the source.
1. Assume that the ultimate U.S. crude oil production will be 200 x 10^9 bbl. Estimate the percentage already produced. Estimate the amount of oil still in place in the U.S. (Note Figures 13 and 14).
2. An oil well is drilled to an 11,000-foot depth. Assume that it produces the average amount of oil per foot drilled and that the oil is sold for $12/bbl. Estimate the total value of the oil obtained from the well.
3. Considering the U.S., determine the relative amount of energy in the natural gas estimated to be recoverable as compared with the crude oil estimated to be recoverable.
4. Estimate (in metric tons) the remaining coal resources of the United States at this time.
5. How long could oil in place in shales in the United States meet the current petroleum production rate.
6. Calculate the cost of 10^6 Btu of energy from oil at $12/bbl and uranium oxide at $10/lb.
7. Discuss the advantages and disadvantages of the use of solar energy as compared with oil to provide a) energy for automobiles, and b) energy for space heating in homes and factories.
8. Would you recommend placing a high tax on crude oil production to conserve this valuable resource? Defend your answer.

Chapter 3

Mechanics, Thermodynamics, and Energy

INTRODUCTION

Energy is a concept—that is, an invention of the mind of man. It is directly associated with motion. We might say it is "that which makes things go," or better perhaps, "something that we use to predict and explain how things go." The power inherent in conceptualization is that one may manipulate concepts, that is, perform mental experiments. The objective is to find agreement between the conclusions reached from these experiments, and those things which we and others observe. The most useful are generalized concepts, which can be used to explain many perceptions of mankind. Such is the concept of energy.

How does one apply broad generalized concepts? The most useful method is to express them in mathematical terms through basic equations, or laws. Concepts, and the laws derived therefrom, form the basis for any science. A formalism is developed that permits one to go either by deductive logic from generalized concepts and laws to specific observations and predictions, or by inductive logic from observed performance to generalizations of such behavior. This is not easy, and not something that can be learned in a short time, particularly when one is dealing with such a powerful and far-reaching concept as energy.

Professor William Reynolds[1] points out that there are three important aspects of energy, which he states as follows:

1. "All matter and all things have energy."
2. "The energy of the whole is the sum of the energy of the parts."
3. "Energy is conserved."

You almost certainly know these three things about energy, though perhaps they have not been so neatly stated. The total energy of a grandfather clock would be the sum of the "swinging" energy of the pendulum, the "wound" energy of the springs, and the "internal" energy stored in all the wood, glass, and metal that make up the clock.

The development of the concept of energy was surely directly associated with the idea that energy is always conserved. It arose first in the study of mechanics, where it is found that if one invents the function called energy certain very difficult problems can be easily solved. Such solutions, however, were limited to a special class of problems, namely those involving conservative force fields. Only when changes were brought about by forces of gravity, electrical charges, elastic springs, or elastic collisions would energy methods yield correct answers. When viscous forces involving friction or inelastic collisions occurred, the methods failed. It took the development of the science of thermodynamics to extend the concept of energy and the law of conservation of energy, so that they could be applied with equal success to all processes, whether or not they involved friction.

In the case of the grandfather clock, it is easy to invoke the law of energy conservation to the pendulum to explain the relation between its velocity at the midpoint of swing to the height it reaches at the end of the swing. This is the familiar conversion of kinetic energy of motion to potential energy in the earth's gravity field and vice versa. But eventually the clock runs down and has to be rewound. Is energy still conserved? Yes, but now it is harder to understand and explain why. The elastic energy stored in the wound metal spring, and the energy of the pendulum due to its motion have been transferred to increased internal energy of the clock and its "surroundings." The energy is now at a molecular level, not readily seen or measured, but nevertheless present. The accounting for this kind of energy transfer is in the realm of thermodynamics, and the generalized concept of energy conservation.

What delayed the acceptance of a general law of conservation of energy was the confusion the early scientists had with heat and the conversion of heat to work, and vice versa. Out of the studies made to resolve this confusion came a second concept in thermodynamics—that of entropy. The consequences of this discovery are enormous and must surely mark the dawn of our modern technological society. That energy is always conserved forms the basis of the First Law of Thermodynamics. The Second Law of Thermodynamics provides a means of determining if a process is possible. Processes which produce entropy are possible—those which destroy entropy are not.

It will be the first objective of this chapter to set forth the basic relationships used by science and technology to study energy and its relationship to physical phenomena. The second objective will be to gain some small experience with the laws of mechanics and thermodynamics, as they are used to determine how transformations of energy from one form to another take place. A principal objective of modern technology is to gain the maximum utilization of energy resources. In particular this applies to the production of work energy from heat energy. Examples are the generation of electric power from the burning of coal, and electricity for space vehicles from the collection of sunlight on solar cells. It is the Second Law of Thermodynamics that determines how efficiently those processes of energy conversion can be performed.

Not all the chemical energy of the coal can be converted to electrical energy. The best efficiencies for coal-fired steam-electric power stations of utility companies is about 40%. This means that 60% of the chemical energy of the coal is not converted to electrical energy, but is rejected to the environment as heat. In common terminology, it is wasted. For the best solar cells used in the NASA space station, or the Bell System orbiting communication satellites, the efficiency is about 16%. Only 16% of the radiated sun's energy striking the solar cell face is converted to electricity; the remainder is reflected or reradiated to space. We shall try, in what follows, to show how energy methods and the concepts of energy and entropy can be used to explain and predict these limitations in efficiency.

ENERGY METHODS IN MECHANICS

Law of the Lever

Let us begin our consideration of energy methods by calling upon one of the simplest possible devices, the lever. The Law of the Lever is stated in terms of two very fundamental concepts, force and distance. Referring to the symbols in Figure 21 the Law of the Lever can be written in mathematical terms as

$$F_1 \cdot X_1 + F_2 \cdot X_2 = 0 \qquad (1)$$

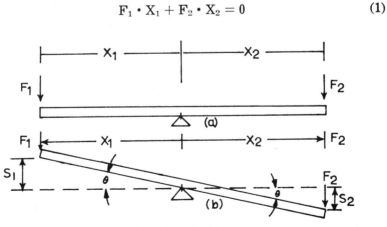

Figure 21. The lever.

In Equation (1) the values of distances X_1 and X_2 have opposite signs, one being measured to the left and one to the right. A way of stating the law is to say: "The *moments* of the forces F_1 and F_2 acting on a balanced frictionless weightless lever are equal but opposite." The moment of a force is the product of the force and its perpendicular distance from a given point.

Let us imagine now that the lever is allowed to rotate through a small angle θ, as in Figure 21b. Multiplying both terms of Equation (1) by θ, there results

$$F_1 \cdot X_1 \cdot \theta + F_2 \cdot X_2 \cdot \theta = 0$$

When the angle θ is small the distance moved at the point of application of the force F_1 is $S_1 = X_1 \cdot \theta$, and similarly setting $S_2 = -X_2 \cdot \theta$. Thus

$$F_1 \cdot S_1 - F_2 \cdot S_2 = 0 \qquad (2)$$

Let us agree to call the product $F \cdot S$ the increment of *work* done. The Law of the Lever could then be stated as: "Two forces acting on a frictionless weightless lever are in balance if the work output by one, during a small displacement, is equal and opposite to the work input of the other."[2] (Note that if work is recognized as a form of energy transfer, then inherent in the formulation of Equation (2) is the principle of conservation of energy.)

Either Equation (1) or Equation (2) is a valid statement of the Law of the Lever. In making a choice between the two in a particular problem solution, experience may dictate one is simpler to apply or learn. The main difference is that the second relies on a more abstract concept, that of "work." The ideas of "force" and "distance" are more familiar, more easily accepted as a result of every-day experience. We might say intuitively they are more basic than work. However, in science we often find that our intuitive ideas are not so productive as abstract thought. If we choose work as a more fundamental concept, many mechanical problems can be solved more easily.

As an example[2] of the advantage of the work concept consider the mechanical linkage in Figure 22. We ask what force F_1 at point *a* will balance a force F_2 at point *d*? If we permit a small displacement upward at point *a* of δ then the displacement at point *d* is 4δ.* For there to be zero net work requires that

$$F_2 \cdot 4\delta + F_1 \cdot \delta = 0$$
$$\text{or} \quad F_2 = -F_1/4$$

---

*The reader should be satisfied that the displacement upward at point b is 2δ and at point c is 3δ.

the negative sign serving only to indicate that the forces are in opposite directions. While the moments of the forces could also be used
to solve this problem, the use of the work concept leads to a much
simpler and direct solution.

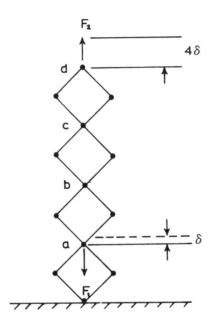

Figure 22. A linkage.

With the concept of work, we can further expand the energy method to the compression of a spring. In Figure 23 we imagine a spring
S being shortened a small amount δ under the action of force F. We
assume the spring displacement small so that the value of the force
does not change. Then, by the principle of conservation of energy
we could write

$$W = F \cdot \delta = \Delta E_{spring} \qquad (3)$$

where

$$\Delta E_{spring} = (E_{final} - E_{initial})_{spring}$$

This relationship states that the work done by the force results in an
increase of energy stored in the spring, the magnitude of the increase
is stored energy, ΔE_{spring}, being precisely equivalent to the work performed.

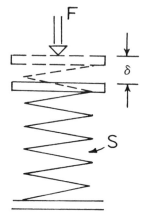

Figure 23. Compression of a spring.

Newton's Second Law

The concept of energy as a conserved quantity began late in the seventeenth century. Its first application was apparently to a freely falling body. To develop the conservation of energy principle for such bodies we began with Newton's Second Law of Motion.

$$F = ma \qquad (4)$$

The equation tells us that a body of mass m, acted on by an unbalanced force F, will increase its speed; the rate of acceleration being a. The velocity continues to increase as long as there is an unbalanced force. When the velocity changes an amount ΔV in a time period Δt then the acceleration is

$$a = \frac{\Delta V}{\Delta t} \qquad (5)$$

This is a good point at which to examine the units assigned to physical quantities in the above relationships.

Letter Symbol	Physical Quantity	Units as assigned in the system		
		British	Physical	International
s	distance	feet (ft)	centimeter (cm)	meter (m)
t	time	second (sec)	second (sec)	second (sec)
V	velocity	(ft/sec)	(cm/sec)	(m/sec)
a	acceleration	(ft/sec^2)	(cm/sec^2)	(m/sec^2)
m	mass	slug (sg)	gram (g)	kilogram (kg)
F	force	pound (lbf)	dyne (d)	Newton (N)

For Equation (4) to hold, one is not at liberty to choose independently the units of force, mass, and acceleration. In the British system, the force unit is chosen as the pound. The mass unit is then defined as that which will be given an acceleration of 1 ft/sec^2 by a force of 1 lbf. It is called a slug. In both the Physical system of units and the International system of units (abbreviated SI units), the mass is chosen as a fundamental unit. In the SI system it is the kilogram (1000 grams). A force of 1 N is then defined as that required to accelerate 1 kg of mass at a rate of 1m/sec^2. Similarly, in the physical system, 1 d of force gives an acceleration of 1 cm/sec^2 to a mass of 1 gram.

It should now be clear that mass is another concept, and is conceived as a property of a body that characterizes its resistance to a change in velocity. Now in everyday language we speak of a person or things weight and express it in pounds. Is weight the same as mass? No, it is a force; the force of attraction by the earth for the object whose weight is stated. Using Newton's Second Law we can write the expression between weight and mass for an object as

$$w = mg \qquad (6)$$

The quantity g is called the "acceleration of gravity." Near the earth's surface its value is very nearly constant so that

$$g = 32.2 \text{ ft/sec}^2 = 9.8 \text{ m/sec}^2$$

From Equation (6) and the value of g we may now deduce that, in the British system of units a slug of mass would weigh 32.2 lb near the earth's surface.

The confusion between mass and weight comes from the fact that we commonly use the same unit to denote both, namely, the pound. Let us rewrite Newton's Law once again, this time as

$$F = ma/g_c \qquad (7)$$

where

$$g_c = 32.2 \text{ ft}-\text{lbm}/\text{lbf}-\text{sec}^2$$

By use of this proportionality constant in Newton's equation, we arrange to have the weight of a body equal in magnitude to its mass near the earth's surface. That is, by Equation (7) the weight or force exerted in the earth's gravitational field is

$$w = mg/g_c \qquad (8)$$

On the earth the acceleration of gravity g has the same value as the proportionality constant g_c, 32.2, hence one pound force (lbf) equals one pound mass (lbm). If both these quantities are abbreviated lb, rather than as lbf for force and lbm for mass, the confusion is increased.

Note carefully the difference in g and g_c. The former is the "local" acceleration of gravity, while the latter is a proportionality constant.

The units of the two quantities are not the same. The mass of a body is invariant with position, whereas the weight depends upon the force of attraction from another body, usually the earth. An astronaut in an orbiting space station is weightless, but his mass is unchanged from that on the earth's surface. A force of one pound, provided by a small gas jet, would give him the same acceleration in the space-craft as it would on the earth, were he on a sled on ice with negligible friction.

Potential and Kinetic Energy

To raise an object in the earth's gravitational field requires an input of energy, namely work energy. Work is the product of force and distance. The force is the object's weight. As has been stated several times, weight of an object near the earth's surface is constant. If then we raise the object from height h_1 to h_2, the work required is

$$W_g = w \ (h_2 - h_2) \qquad (9a)$$

or by Newton's Second Law

$$W_g = mg \ (h_2 - h_1) \qquad (9b)$$

or

$$W_g = (mg/g_c) \ (h_2 - h_1) \qquad (9c)$$

In Equation (9b), if height change is in feet then mass is in slugs, whereas in Equation (9c), a height change in feet is accompanied by mass in pounds, lbm. In both cases the work, W, would then be expressed in ft-lbf.

If the mass m is moved without acceleration, then the upward force required to lift it through the height change is the same as its weight. This is illustrated in Figure 24a. In this case the work done by the lifting force is equal and opposite to the work done by the gravity of the earth. Designating the latter by W_g we have

$$W_g = -w(h_2 - h_1) = w(h_1 - h_2) \qquad (10)$$

or

$$W_g = PE_1 - PE_2 = -\Delta PE \qquad (11)$$

where $\Delta PE = w(h_2 - h_1)$ and $PE = wh = m(g/g_c)h$ $\qquad$ (12)

Equation (11) can be stated as: "The net work done by the force of gravity on a body is equal to the decrease in its potential energy."

It is appropriate to ask, where is this potential energy after the weight has been lifted? Is it in the mass m, or in the earth? The answer can be illustrated through Figure 24b. We might imagine a

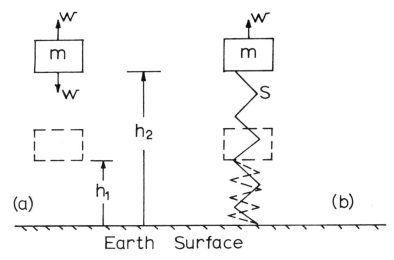

Figure 24. Illustrating potential energy increase is analogous to stretching a spring.

massless invisible spring stretched between the earth and object exerting a constant downward force w. If this spring is extended by slow upward movement from a balancing force w, energy will be stored in the spring equal to the work done. This was earlier illustrated in Figure 23 and Equation (3).

Potential energy then is neither stored in the body moved nor in the earth, rather it is stored in the conservative force field between them. Nevertheless, we find it convenient to associate the potential energy with the object. We say a boulder at the top of a hill has potential energy, and we can recover that energy and convert it to work, by lowering it to the bottom against some counterbalancing force.

Combining Equations (4) and (5) leads to

$$F = m\,\frac{\Delta V}{\Delta t}$$

If we apply a constant force F to a body of fixed mass m, then the velocity increase ΔV is directly proportional to the increase of time. Such a relationship is plotted in Figure 25. The velocity increases linearly with time, *i.e.*, the plot of V versus t is a straight line. If the velocity starts at zero and goes to a value V in time t, the average velocity is

$$V_{av} = \tfrac{1}{2}V$$

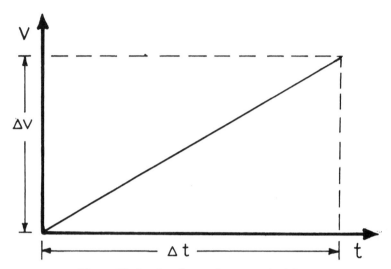

Figure 25. Acceleration under a constant force.

The distance covered in this time period is

$$S = V_{av}t = \tfrac{1}{2}Vt$$

The force providing this acceleration is

$$F = mV/t$$

The work done by the force over time period t is then

$$W = F \cdot S = \tfrac{1}{2}mV^2 \qquad (13)$$

or

$$W = \Delta KE = KE \qquad (14)$$

where

$$KE = \tfrac{1}{2}mV^2 \qquad (15)$$

Equation (14) can be stated as: "The net work done by a force on a moving object is equal to the increase in its kinetic energy." The object has energy by virtue of its translational motion. We call it kinetic energy. The value is computed from Equation (15). If the mass of the object is expressed in pounds (lbm) and velocity in ft/sec, then the expression for kinetic energy must be written as

$$KE = mV^2/2g_c \qquad (16)$$

When a body is acted on by the force of gravity it will have potential energy by virtue of its position above (or below) the earth's surface, and can also have kinetic energy, if it is in motion relative to the earth. From Equation (11) and Equation (14), we can write for the work done by the earth's gravitational field

$$W_g = PE_1 - PE_2 = -\Delta PE$$

and

$$W_g = KE_2 - KE_1 = \Delta KE$$

Combining these two equations gives

$$PE_1 + KE_1 = PE_2 + KE_2 \qquad (17)$$

or $\qquad E = PE + KE = \text{constant} \qquad (18)$

This is the principle of conservation of energy applied to a conservative force field. It is restricted to two forms of "stored" energy, kinetic energy of translational motion and potential energy of position in a gravity field. The derivation of the foregoing equations has been based on constant forces applied to a body. This was done to avoid use of the calculus. Readers familiar with calculus will recognize this restriction of constant forces is not required.

Example 1

Consider the pendulum of a grandfather clock with a mass of 1.6 lbf. At the end of its swing it is 0.20 inches above the center position. What velocity will it have as it crosses the center position.

Designate by 1 the position at the end of the swing and by 2 the center position. Let the potential energy be measured relative to the pendulum position at the center. Then

$$PE_2 = KE_1 = 0$$

From Equation (17)

$$PE_1 = KE_2$$

and from Equations (12) and (16)

$$m(g/g_c)h_1 = mV_2^2/2g_c$$

so

$$V_2^2 = 2gh_1 = 2(32.2)\,(0.20/12)$$

$$V_2^2 = 1.07 \text{ ft}^2/\text{sec}^2$$

$$V_2 = \sqrt{1.07} = 1.036 \text{ ft/sec}$$

WORK

Generalized Concept of Work

Work in a purely mechanical system can always be identified as a force acting through a distance, *i.e.*, in equation form as $F \cdot S$ for a small displacement S. The concepts of work and energy are, as we know, useful for systems other than in mechanics. Examples are the work of hot gases expanding against a piston in an automobile engine, work of a battery in starting a car, work output of a photovoltaic solar cell, and of a pneumatic drill. In each of these examples the net effect of the work is equivalent to a force acting through a distance. It is, in fact, possible in every case where work is done to express it in equation form as

$$W = F_k \cdot X_k \tag{19}$$

where $F_k =$ a generalized force

$X_k =$ a generalized displacement

Expansion and Compression of Gases

A gas contained in a cylinder of fixed cross-sectional area (A) and closed by a piston, is illustrated in Figure 26. For a small displacement of the piston δ the increment of work done by the gas will be

$$W = F \cdot \delta = pA\delta = p\Delta v \tag{20}$$

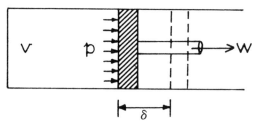

Figure 26. Work by an expanding gas.

Comparison of Equations (19) and (20) indicates that in the case of gas expansion the generalized force is gas pressure and the generalized displacement is volume change of the gas.

Consider now a process, like that in Figure 26, where the gas changes from an initial condition of the pressure p_1 and volume v_1 to pressure p_2 and volume v_2. This change in the state or condition of the gas might be accomplished in several different ways. Three possible paths of the course of change in pressure and volume from state 1 to state 2 are shown by the lines labeled (a), (b), and (c) in Figure 27.

Along path (a) the work done by the gas on the piston is

$$W_a = p_1(v_2 - v_1)$$

It may be seen that this corresponds to a certain area on the pres-

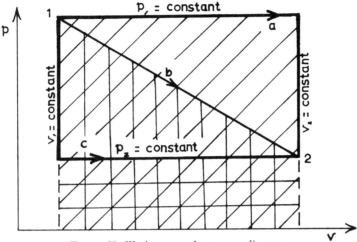

Figure 27. Work areas of an expanding gas.

sure-volume plot of Figure 27. That area is designated by diagonal line shading.

Along path (c) the work done by the gas is

$$W_c = p_2(v_2 - v_1)$$

Obviously this is a smaller amount of work than for path (a). The area designated by horizontal line shading represents the work W_b in Figure 27. The vertical line shading in the figure depicts the work done by the gas along path (b). The work energy by W_b will be less than W_a but greater than W_c.

From the above example we draw a very important conclusion about the quantity work. *Work depends on the process path.* Work is not a property of a body or system. Rather it is an energy *transfer* and depends upon the manner in which a given change in a system takes place. When a system exchanges work energy with its surroundings the process can always be identified as equivalent to the raising or lowering of a weight, *i.e.*, a force acting through a distance.

Electrical Work

One of the most useful and common forms of work is by a change of state of electrical charge. Consider the battery and starting motor of an automobile as illustrated in Figure 28.

The unit of electrical charge is the coulomb (q) and the electromotive force is the volt (ϵ). When a small charge taken from the battery passes to the motor the work performed is

$$W = \epsilon \cdot q \tag{21}$$

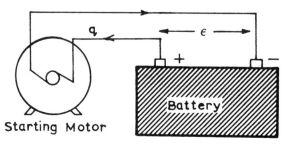

Figure 28. Automobile battery and starting motor.

and the work is expressed in joules. A work *rate* of 1 joule per second is a watt. In Equation (21) the voltage (ϵ) is the generalized force and the charge (q) is the generalized displacement. The movement of the charge by the electromotive force constitutes work. The effect is readily observable as an energy transfer and is equivalent to the raising of a weight. If the motor was an ideal one (*i.e.*, without losses), it was attached to a weight by a frictionless pulley and rope arrangement, the weight lifted one foot would be equal to the identity between a joule and a foot-pound, namely

$$1 \text{ ft-lbf} = 1.356 \text{ joules} = 1.356 \text{ watt-seconds}$$

or   1 watt-hour $= 2.655$ ft-lbf

and   1 kilowatt-hour $= 2.655 \times 10^6$ ft-lbf

HEAT

There is another way in which energy can be transferred to a system, one that is not observable as work. This second type of energy transfer is called heat. Heat is very closely associated with the concept of *temperature*. Both heat and temperature are concepts central to the subject of thermodynamics. When a body at a higher temperature is brought into contact with one at a lower temperature, energy passes from the hotter to the colder body. That energy transfer is heat. A temperature difference can be viewed as the driving force for heat transfer.

To understand the difference between heat and work as modes of energy transfer, it is helpful to think of how they take place on a molecular level. Imagine a gas in a cylinder closed by a piston, as in Figure 26. If the piston is moved inward work is performed on the gas. The molecules of the gas gain energy, but in a very directed sense, *i.e.*, in an organized manner. Let the piston now be fixed in position, and imagine that the boundary walls of the cylinder are made up of closely packed atoms that are vibrating. Some will be moving toward the gas, some parallel to it, and others at various angles to the boundary. Their interactions with the gas molecules can result in a transfer of energy. If the wall atoms are caused to increase their vibrational rate, additional energy transfer will take place. This transfer of energy is heat. It is a disorganized type of energy transfer.

As one molecule "sees" an energy transfer, it is not possible to tell if it is due to a heat or work effect. Work is always recognizable on a macroscopic or human perception scale. Heat is not; it is required on the macroscopic scale, however, to account for disorganized energy transfer.

It is the custom to use the symbol Q for heat. Before the development of the science of thermodynamics heat was thought of as a substance. It was called "caloric," and when heat was added to a substance its caloric content was increased. Unfortunately we still find remnants of this idea in such terms as the "heat capacity" or "specific heat" of a substance. In fact heat is neither stored in a system nor is its heat content increased. When heat is transferred to a body, the energy of that body increases. The amount of energy transferred as heat, when a system changes from one condition to another, depends *on the process path*. The term "heat content of a body" is meaningless, just as is "work content of a body." Energy is what the body contains, and we can change its energy content by either work or heat interactions with other bodies.

In the British system the unit of heat energy is the British thermal unit or Btu. It is the energy required to raise one pound mass of water 1°F (or more precisely from 59.5°F to 60.5°F). In the Physical system the unit of heat energy is the calorie or kilocalorie. One calorie is the energy required to raise the temperature of one gram of water 1°C (or more precisely from 14.5°C to 15.5°C). In the SI or International system the standard unit of energy is the joule or its equivalent the watt-second. The relationship between these quantities and other energy units are as follows:

$$
\begin{aligned}
1 \text{ Btu} &= 1055 \text{ joules} = 778 \text{ ft-lbf} \\
1 \text{ cal} &= 4.187 \text{ joules} = 3.088 \text{ ft-lbf} \\
1 \text{ Btu} &= 252 \text{ calories} = 0.252 \text{ kcal} \\
1 \text{ Btu} &= 2.93 \times 10^{-4} \text{ kwh} \\
1 \text{ watt} &= 3.413 \text{ Btu/hr} = 0.7376 \text{ ft-lbf/sec}
\end{aligned}
$$

MOLECULAR ENERGY

We have seen that on a macroscopic scale the term energy can be associated with motion of a body. The term $[mV^2/2g_c]$ is called the kinetic energy. It is a property of the body and is dependent upon its mass and velocity. If a faster moving body collides with a slower moving one it may increase the speed of the slower one, while decreasing its own speed. A kinetic energy transfer takes place. This same kind of energy transfer can occur at the molecular level. Matter is composed of molecules and the molecules have atoms with a nucleus and electrons. All are in motion. There are translational, rotational (spin), and vibrational motions. Potential energies at the molecular level are also present with binding forces between nuclei, binding forces between atoms and molecules. There is coulomb binding between charged particles like protons and electrons, and electric and magnetic dipole moments due to magnetic and electric fields.

As energy is added to a substance some of the molecules move to more excited states. This is characterized by faster motions, increased binding energies, etc. of the microscopic particles which make up the substance. The energy associated with these many forms of molecular activity we call the *internal* energy of the substance. If we add this energy to the "bulk" energy modes derived earlier, we may express the total energy of a substance as

$$E = KE + PE + U \tag{22}$$

where E = total energy of the substance

and U = internal energy of the microscopic modes

FIRST LAW OF THERMODYNAMICS

The First Law of Thermodynamics is an extension of the conservation of energy principle in mechanics to include all physical systems of fixed mass, having both work and heat interactions with its surroundings. We distinguish between these two types of energy transfers, work and heat, and write the general law of energy conservation for a system of fixed mass as

$$(W + Q)_{in} = \Delta E \tag{23}$$

or

$$W_{in} + Q_{in} = (U_f + KE_f + PE_f) - (U_i + KE_i + PE_i) \tag{24}$$

The equation states that the work and/or heat added, or put in to a system, is equal to the increase of energy of that system. The energy increase is the energy content at the condition or state of the system after the energy transfers were made, less the energy content at the state or condition of the system before those transfers. Those energy contents consist of the molecular internal energy, kinetic energy of bulk motion, and bulk potential energy in the gravitational field. If the work is performed by the system, W_{in} would be negative, or $W_{in} = -W_{out}$. Similarly heat transfer out of the system is related to heat input, by, $Q_{in} = -Q_{out}$.

It should be clear that if no energy transfer processes are permitted the energy of a fixed mass or "closed" system is a constant.

That is

$$\Delta E_{isolated} = 0$$

or

$$E_{isolated} = constant$$

(25)

A system can always be chosen large enough so that no energy crosses the system boundaries. For such a system, Equation (25) restricts all possible changes in the state to those having the same total energy content.

There is no proof for the above equations, except to say that in the millions of applications of the First Law of Thermodynamics to predict or explain physical phenomena, the results have always been satisfactory. The relevant question about the energy concept and its logical extension of conservation of energy is not whether it is true, but what can be done with it? The answer is that almost all science and engineering analyses involve considerations of an energy balance.

Consider a conventional nuclear power plant as illustrated in Figure 29. In the reactor, fission of U^{235} results in an energy

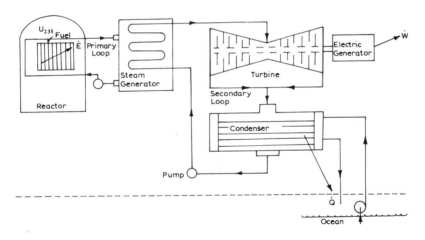

Figure 29. Nuclear steam power plant.

release rate $\dot{E}$ which is transferred to high-pressure water, raising its temperature. The heated pressurized water is pumped through piping of a primary loop to the steam generator. Here it transfers energy to water in a secondary loop, producing high-temperature, high-pressure steam. The steam is delivered to a turbine where it is expanded through nozzles to high velocities, which impart energy to moving blades. The turbine work produced drives an electric generator from which emerges a work output rate $\dot{W}$. The exhaust steam of the turbine passes to a condenser where low-temperature seawater is circulated. Energy is extracted by the sea water at a rate $\dot{Q}$ causing the steam to condense. The secondary loop is completed by a pump which takes the low-pressure water from the condenser and returns it at a high pressure to the steam generator.

If the required electric generating capacity is 100 Mw (megawatts = 10^6 watts), and the overall plant efficiency is 33%, then

$$\eta = \dot{W}/(-\dot{E}) = 0.33$$

and $$\dot{E} = -100/0.33 = -300 \text{ Mw}$$

For each 100 Mw of work rate out, the uranium gives up or reduces its energy at the rate of 300 Mw.

Let us choose as a thermodynamic system all matter that is enclosed within the dashed line of Figure 29. We apply the First Law of Thermodynamics to this system of fixed mass and write

$$W_{in} + Q_{in} = \Delta E$$

or alternatively

$$W_{out} + Q_{out} = -\Delta E$$

We can divide each of these quantities by some time interval Δt so that

$$\left(\frac{W}{\Delta t}\right)_{out} + \left(\frac{Q}{\Delta t}\right)_{out} = -\frac{\Delta E}{\Delta t}$$

or $$\dot{W}_{out} + \dot{Q}_{out} = -\dot{E}$$

then $$\dot{Q}_{out} = -\dot{W}_{out} - \dot{E}$$

The rate of heat rejection to the seawater in the condenser is therefore

$$\dot{Q} = -100 - (-300) = 200 \text{ Mw}$$

Of the 300-Mw energy rate taken from the uranium fuel elements, 100 Mw are converted to useful electrical energy and 200 Mw are discarded to the ocean. Why such an outlandish energy rejection? An efficiency of 33% is typical for modern nuclear power plants. To understand why requires that we examine the Second Law of Thermodynamics and the characteristics of power generating stations.

SECOND LAW OF THERMODYNAMICS

Irreversible Processes

If energy is always conserved, as required by the First Law of Thermodynamics, then why do we seem to be "running out" of energy? There seems to be something missing that will explain why nature selects certain directions for energy transfers to take place.

We are aware every day of this directional character in the things we observe. A ball of putty dropped from a few feet above the floor will convert its potential energy to internal energy, *i.e.*, increased energy levels of the putty molecules. No amount of coaxing will cause the molecules to reduce their energy and "lift" the ball into the air again. The kinetic energy of a moving automobile can be transferred to internal energy of the brake drums as it is brought to a quick stop. There is no way to reverse that process. Work from a small electric motor can act through a stirrer to raise the temperature of water in a container. While with some elaborate efforts it may be possible to get some of that stored energy in the water back as work, only a small fraction could be so converted. Hydrogen and oxygen as a gas mixture can be ignited with a small spark to form water, and in the process, perform work or produce high temperatures. The water, so formed, cannot be separated back to hydrogen and oxygen without much more work being done.

In each of the above illustrations, the First Law would not be violated if the energy transfer process were reversed. But they are irreversible processes. A concept is needed that will permit an explanation of the directionality of physical processes, and a law derived therefrom, to predict the maximum utilization of energy through those processes that are possible. Such is the concept of *entropy*, a property that can be thought of as a quantitative measure of the microscopic *randomness* of the energy distribution among the molecules. Thermodynamic entropy can be directly related to a scale of *uncertainty* as it would be used in information theory.[3] In the case of thermodynamics it is a measure of our uncertainty as to how the total energy of a system is distributed among the multitude of quantum molecular energy levels available.

The directionality of real processes is from states of lower entropy to ones of higher entropy, *i.e.*, to greater randomness of energy distribution at the microscopic level, and hence to greater uncertainty about the distribution. Processes which result in an increase of entropy are irreversible and possible, those in which the entropy is unchanged are reversible, and those for which an entropy decrease would occur are impossible.

A little thought about each of the examples of irreversible processes given above will reveal that they result in changing a directed form of energy content to a more random energy content. In the one case of the combustion of hydrogen this may be more difficult to conceive.

The Entropy Concept

As in the case of energy transfer by heat, it is helpful to visualize what is happening at the molecular level when attempting to "explain" what is meant by an entropy increase. Consider a box containing a gas and imagine the gas made up of many molecules that are more or less like hard rigid spheres in unceasing motion. An equilibrium condition exists, that is, there are no pressure or temperature gradients in the gas. Suppose we insert a partition down the center of the box and imagine we could make the color of all the molecules on the left side red, and those on the right side white. When the partition is removed diffusion will occur with red particles moving from left to right and white particles from right to left. Viewing the box from some distance would give the effect of a gradual change in color to pink. At *equilibrium* there would be an equal distribution of red and white particles on both sides of the box, a state of maximum randomness of distribution would be reached.

The above analogy may serve to illustrate the nature of the entropy concept in thermodynamics. Entropy, like energy, is an extensive property of matter, that is, the entropy of the whole is the sum of the entropy of the parts. Consider again the box with two compartments, gas A on one side, and gas B on the other. The internal energy of the contents of the box is the sum: $U_C = U_A + U_B$. Similarly, the total entropy of the box contents is the sum: $S_C = S_A + S_B$. There is a very significant difference in these two quantities, however. If the box is isolated, *i.e.*, no heat or work interactions, the energy content must remain constant. The entropy of the isolated box can either remain constant or increase, but never decrease. Processes resulting in an entropy increase go toward equilibrium states which are characterized by maximum randomness or greatest uncertainty about microscopic energy distribution.

The Second Law of Thermodynamics can be stated as

$$\Delta S_{isolated} > 0 \tag{26}$$

This equation should be compared with Equation (25) for energy of an isolated system.

HEAT ENGINES

Efficiency of a Heat Engine

While the above discussion gives us some ideas of the nature of entropy, we need some means of applying it to determine how conversion of energy from one form to another can be made. In particular, we wish to know how to convert thermal energy such as heat from the combustion of fuels, nuclear reactions, or solar radiation to work energy. Electrical energy is one of the most useful forms of energy.

Figure 29 illustrated a power plant for the generation of electrical energy. The general name for such plants is *heat engine*. We can make the fundamental postulate for the Second Law in terms of a heat engine, namely:

Kelvin-Planck Statement

"It is impossible to construct a device which will operate *continuously* and have no effect other than to produce work and extract heat from a single source at a uniform temperature."

Figure 30 again illustrates the major equipment items in a steam generation power plant, and the *cyclic* process which the water, as a working fluid, undergoes. The cyclic process is represented on a plot of pressure versus volume. The steam at a high pressure (P_{boiler}) expands in the turbine (process 2-3) performing work against moving blades, as explained previously, and leaves at a low pressure ($P_{condenser}$). The volume of steam increases enormously in this process. Typically, a pound of steam at the turbine inlet will have a volume of less than one-half cubic foot, whereas at the turbine exhaust that pound of steam would have expanded to a volume of more than 1000 ft³. In the condenser (process 3-4), heat is removed from the low-pressure steam by cooling water, and liquid water is returned to a boiler feed pump. The pump raises the water again to boiler pressure at, essentially, constant volume, since water is nearly incompressible. In the boiler or steam generator (process 1-2), the water is evaporated and superheated to again produce the high-temperature steam at state 1 delivered to the turbine completing the cycle.

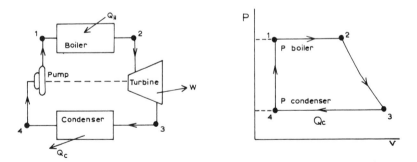

Figure 30. Heat engine and its cyclic process shown on a pressure-volume diagram.

There are two major places of heat transfer in the cycle described. First, is a heat addition, Q_H, which comes from the burning of coal, oil, or gas; or, in the case of a nuclear plant, the energy release of fissioning uranium. Second is a heat rejection, Q_c, made to the cooling water from a river, lake, cooling tower, or the ocean.

The Kelvin-Planck statement of the Second Law requires that there be a heat rejection Q_c for the heat engine of Figure 30. The efficiency of the heat engine is defined as

$$\eta = \frac{\text{Net Work Output}}{\text{Heat Addition}} = \frac{W}{Q_H} \tag{27}$$

In a cyclic process, the working fluid continuously returns to the same state, hence has no change in energy content for an integral number of cycles. As a consequence, application of the First Law to a heat engine cycle, gives

$$W = Q_H - Q_C \tag{28}$$

and, consequently,

$$\eta = \frac{Q_H - Q_C}{Q_H} = 1 - \frac{Q_C}{Q_H} \tag{29}$$

So long as the heat rejection, Q_C, is finite (which it must be) the heat engine efficiency is less than 100%.

Absolute Temperature and its Relationship to Heat Engine Efficiency

Let us determine now what efficiencies are possible from a heat engine. To do this we conceive of a *reversible* heat engine, one which can operate in either of the following modes: a) Heat Engine—takes heat from a high temperature source, produces net work, and rejects heat to a lower temperature sink; or b) Heat Pump—extracts heat from a low-temperature source, uses net work, and rejects heat to a higher temperature sink. An example of a heat pump is a household refrigerator.

Consider now the situation illustrated in Figure 31. Two reversible heat engines operate between a high-temperature and low-temperature reservoir at uniform temperatures of T_1 and T_2 respectively. Let us suppose that one of these heat engines has a higher efficiency than the other, say A more efficient than B. If so, then for the same heat input Q_1, to each engine

$$W_A > W_B$$

then

$$\eta_A = \frac{W_A}{Q_1} > \eta_B = \frac{W_B}{Q_1}$$

Since both heat engines are reversible we postulate operating B as a heat pump. As $W_A > W_B$ we can drive the heat pump B by heat engine A, and still have some net work left over, W_{NET}. Now since Q_1 is the same for heat engine A, and the "reversed" heat engine B we can eliminate the need for the Hi-temp reservoir and simply direct the heat output from the pump to the engine. This is indicated by the

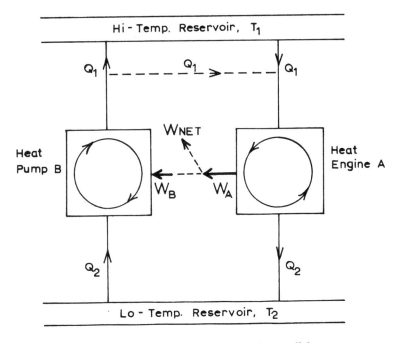

Figure 31. Reversible heat engines in parallel.

dotted line in Figure 31. But this is impossible! Why? Because we can consider engines A and B in combination as a device which operates continuously, produces net work, and exchanges heat with only a single reservoir. (Remember that we eliminated the need for the reservoir at T_1, hence only the Lo-Temp reservoir at T_2 must be used.) Such a device is a direct violation of the Kelvin-Planck Statement of the Second Law.

The necessary conclusion of the postulated experiment above is: "All reversible heat engines operating between two heat reservoirs at fixed and uniform temperatures must have the same efficiency." It is easy to see how this statement can be further expanded to: "No heat engine operating between two reservoirs at fixed and uniform temperatures can exceed in efficiency that of a reversible heat engine." From Equation (29), it further follows that for a reversible heat engine operating between two uniform temperature reservoirs at T_1 and T_2

$$\eta_{REV} = 1 - \frac{Q_2}{Q_1} = f(F_1, T_2) \tag{30}$$

and

$$\left(\frac{Q_1}{Q_2} \right)_{REV} = F(T_1, T_2) \tag{31}$$

That is to say, the efficiency, η, and the ratio of the heat received to the heat rejected, Q_1/Q_2, are functions *only* of the temperatures of the two reservoirs, T_1 and T_2. Since no engine can exceed in efficiency a reversible engine, Equation (30) tells us there is a maximum possible efficiency which is set, once the temperatures of the heat reservoirs are fixed.

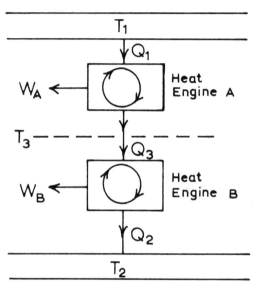

Figure 32. Reversible heat engines in series.

We may carry our logic an additional step by imagining two reversible heat engines in series as illustrated by Figure 32. The heat rejected from engine A, Q_3, is also the heat addition to engine B. Let us designate the temperature of this intermediate region of heat exchange as T_3. Now, from Equation (31) we may write

$$\frac{Q_1}{Q_2} = F(T_1, T_3); \quad \frac{Q_3}{Q_2} = F(T_3, T_2); \quad \frac{Q_1}{Q_2} = F(T_1, T_2)$$

The last equation relates to heat engines A and B, taken together, as a single device operating between heat reservoirs at T_1 and T_2. Now

$$\frac{Q_1}{Q_3} = \frac{Q_1/Q_2}{Q_3/Q_2} \text{ so, } F(T_1, T_3) = \frac{F(T_1, T_2)}{F(T_3, T_2)}$$

All three temperatures are independent, hence T_2 must divide out of the ratio in the final relationship above and

$$\frac{Q_1}{Q_3} = F(T_1, T_3) = \frac{\phi(T_1)}{\phi(T_3)} \tag{32}$$

The choice of the function $\phi(T)$ is arbitrary. A simplest possible choice, and that proposed by Kelvin, which has been universally adopted is

$$\phi(T) = T$$

Thus, by Kelvin's definition, a reversible heat engine operating between two uniform temperature reservoirs is subject to the relationships

$$\frac{T_1}{T_2} = \left(\frac{Q_1}{Q_2}\right)_{REV} \tag{33}$$

$$\eta_{REV} = 1 - \left(\frac{Q_2}{Q_1}\right)_{REV} = 1 - \frac{T_2}{T_1} \tag{34}$$

The first of these two equations establishes an absolute thermodynamic temperature scale, independent of any substance. Of course we usually measure temperature by mercury expansion in a glass capillary tube, or movement of a linkage attached to a bimetallic strip. Scientists working with liquid helium, near the absolute zero on the Kelvin temperature scale, actually measure temperatures by a heat ratio expression as in Equation (33). Further, primary and secondary standard temperatures for the International Temperature Scale are usually determined by very careful measurements with an ideal gas thermometer, which can be directly related to the Kelvin definition of Equation (33).

The relationship between absolute temperature scales and those in common use are as follows:

$$T(°Kelvin) = T(°Celsius) + 273.2$$

$$T(°Rankine) = T(°Fahrenheit) + 459.7$$

On the Kelvin scale the freezing point and normal boiling point of water are 273.2°K and 373.2°K respectively, corresponding to 0°C and 100°C on the Celsius (formerly called centigrade) scale. On the Rankine scale, these same temperatures are 491.7°R (=32°F + 459.7), and 671.7°R (=212°F + 459.7). Conversion from Kelvin to Rankine temperatures is simply made by

$$T(°Kelvin) = (5/9)T(°Rankine)$$

which corresponds to the fact that 100°K, and 180°R separate the normal freezing and boiling points of water.

Absolute temperature scales have a range of zero degrees to infinity. Absolute zero temperature is impossible to attain, though it has been approached "closely" (within a few millionths of one degree) by scientists working in the area of ultralow temperatures. Strange things happen at these ultralow temperatures. One of the

most fascinating and extensively investigated phenomena is superconductivity, the disappearance of resistance to electrical current
flow in a conductor. A current started in a closed loop of a superconductor will flow forever. The consequences for reducing energy
losses in transmission of electrical power make this an exciting prospect of the future.

The relationship we were most interested in seeking here, however,
is Equation (34). It established the maximum efficiency which can
be realized with heat engines operating between two levels of temperature.

Example 2

What efficiency is possible in the conversion of heat to work in a
modern steam-electric generating station? The highest temperature
which can be used in a steam turbine, without frequent replacement
of blading is about 1200°F. Heat rejection to the atmosphere, lakes,
or ocean will be at about 100°F. Let us assume then that all the heat
addition is at $T_1 = 1200°F$ and all of the heat rejection in the condenser at $T_2 = 100°F$. The maximum heat engine efficiency realizable
under these conditions is

$$\eta_{REV} = 1 - \frac{T_2}{T_1} = 1 - \frac{100 + 459.7}{1200 + 459.7} = 0.663$$

An efficiency of 66.3% is possible. However, the very best steam electric generating stations have an overall efficiency of near 40%, while
the average for all electric generation in the USA is slightly less than
30%. Why so much lower than what is possible? There are a number
of practical limitations. In the steam generator not all the heat can
be added at 1200°F; the water must be heated to the boiling point,
then evaporated, and the steam superheated to 1200°F. The turbine,
electric generator, and pump all have rotational friction losses, *i.e.*,
they are not "reversible" machines. There are other reasons some
of which will be explored in the following chapters on power plants.

You might also ask, why not use even higher temperatures than
1200°F? The principal reason is materials limitations. What do you
make turbine blades from that will withstand such high temperatures
for long periods of time, without failure or frequent (and costly) replacement? Well if we can't make T_1 very high, how about reducing
T_2 to a lower value? The heat sink, at T_2, must be capable of absorbing large quantities of heat without a significant change in temperature. This limits us to the atmosphere (usually with cooling towers),
large lakes (natural or artificial), or the oceans, over whose temperature we have no control.

Entropy and the Heat Engine

We have shown that a consequence of the Kelvin-Planck statement of the Second Law of Thermodynamics for a heat engine operating between two uniform temperature heat reservoirs is, after Equation (33)

$$\frac{(Q_1)_{REV}}{T_1} = \frac{(Q_2)_{REV}}{T_2} \tag{35}$$

When such a relationship holds we obtain the maximum efficiency for a heat engine operating between heat reservoirs at T_1 and T_2, *i.e.*, maximum work output is obtained for a given heat transfer. This suggests that (Q_{REV}/T) could be used as a measure of some kind to tell when we are using energy efficiently. In classical thermodynamics a new property of matter, *entropy*, is defined by the relationship

$$\Delta S = \frac{Q_{REV}}{T} \tag{36}$$

If a system receives heat from a number (say n) uniform temperature reservoirs, then the change in entropy is

$$\Delta S = \sum_{i=1}^{n} \frac{(Q_i)_{REV}}{T_1} \tag{37}$$

In the case of a reversible heat engine taking heat Q_1 from a reservoir at T_1, and rejecting heat Q_2 to a reservoir at T_2, the entropy change is

$$\Delta S)_{REV\ HEAT\ ENGINE} = \frac{(Q_1)_{REV}}{T_1} - \frac{(Q_2)_{REV}}{T_2} = 0 \tag{38}$$

which follows from Equation (35). Note that Q must be considered positive when heat is added to a system and negative when it is removed.

Let us repeat here a statement made earlier in introducing the Second Law of Thermodynamics. That statement was: "The directionality of real processes is from states of lower entropy to ones of higher entropy, *i.e.*, to greater randomness of energy distribution at the microscopic level and hence to greater uncertainty about the distribution." It will be difficult for the reader who has not previously encountered thermodynamics to "see" the relationship between entropy change as defined by Equation (36), and the concept of entropy as a measure of randomness in energy distribution at the microscopic level. Nevertheless the relationship does exist and its development can be found in treatises on statistical thermodynamics.[3,4]

Perhaps it will help some to think about the quantity (Q/T) and its effect on a system. Heat is a disorganized type of energy transfer. When heat is added to a system there is an increase in disorganized energy, hence an increase in entropy. An increase in temperature of matter generally means an increase in energy, and greater random-ness of energy distribution among its microscopic modes of internal energy. Consider the same amount of heat transferred to two identi-cal systems, except that one is at a higher temperature than the other. According to Equation (36), the one at higher temperature would have less entropy increase since Q/T_{HI} would be less than Q/T_{LOW}. The system at higher temperature already has a higher state of uncertainty before the heat addition. Consequently the same amount of heat transfer makes a smaller change in the randomness of energy distribution for the system at higher temperature.

ENTROPY AND WORK AVAILABILITY

There are three fundamental aspects of the concept of entropy that parallel those stated for energy on the beginning page of this chapter. They are as follows:
1. All matter has entropy, which is a measure of the uncertainty con-cerning the distribution of energy among the many microscopic modes available.
2. Entropy is a property of matter which is extensive, the sum of the entropy of the parts is equal to the whole.
3. Entropy is created in real, irreversible processes. It can never be destroyed.

The latter statement leads to an alternative expression for the Sec-ond Law, namely

$$\Delta S_{Isolated} > 0 \tag{39}$$

This is the statement of the Second Law comparable to Equation (25) for the First Law, *i.e.*, $\Delta E_{isolated} = 0$.

If some part of the world, be it small or large, is isolated from every other part so that no energy transfer can occur across its boundaries, all subsequent states or configurations of that isolated system must be ones with the same energy content, and with the same or higher values (but never lower) of entropy. States of higher entropy are ones of greater randomness of energy distribution, ones with less availability for the production of work, and ones which are most stable.

Example 3

Suppose that a metal rod is placed between two heat reservoirs one at a temperature of 1000°R and the other at a temperature of 500°R. An amount of heat $Q_1 = 1000$ Btu is transferred through the rod from the upper to the lower reservoir as illustrated in Figure 33. How much entropy increase is created by this process? What availability to produce work was lost?

In answer to the first question, the entropy decrease of the reservoir at T_1 is

$$\Delta S_1 = \frac{Q_1}{T_1}\bigg|_{\text{REV}} = \frac{-1000}{1000} = -1 \text{ Btu/}^\circ\text{R}$$

negative since heat is transferred out of the reservoir. The entropy increase of the second reservoir is

$$\Delta S_2 = \frac{Q_1}{T_2}\bigg|_{\text{REV}} = \frac{+1000}{500} = +2 \text{ Btu/}^\circ\text{R}$$

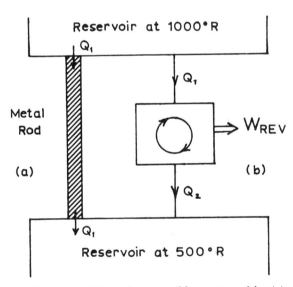

Figure 33. A reversible and irreversible process of heat transfer.

There is no entropy change of the rod since it is in a "steady state," *i.e.*, it has the same temperature distribution at the beginning and end of the process and the same energy content. Then for the entire system of reservoirs and rod

$$\Delta S_{\text{net}} = \Delta S_1 + \Delta S_2 = -1 + 2 = +1 \text{ Btu/}^\circ\text{R}$$

As should be expected from the Second Law there is a net entropy increase since the process is irreversible. In fact an alternative statement of the Second Law called the Clausius statement is: "It is impossible for a system working in a complete cycle to accomplish as its sole effect the transfer of heat from a body at a given temperature to a body at a higher temperature."

For the example here, this means the energy transferred by the rod from the 1000°R to the 500°R reservoir cannot be returned in any cyclic process without an input of work. Let us turn now to the second question concerning the loss of availability to perform work. Suppose the same heat energy, $Q_1 = 1000$ Btu, were taken from the upper resrvoir and delivered to a reversible heat engine as illustrated in Figure 33b. Heat is rejected to the lower reservoir by the engine. The engine efficiency is

$$\eta = \frac{T_1 - T_2}{T_1} = \frac{1000 - 500}{1000} = 0.50$$

The work performed is

$$W_{REV} = \eta Q_1 = 0.50(1000) = 500 \text{ Btu}$$

and

$$Q_2 = Q_1 - W_{REV} = 500 \text{ Btu}$$

You can readily show that, in this case, there is no net gain of entropy for the entire system of reservoirs plus engine.

The reader might be inclined to ask, where did the entropy come from when the metal rod transferred heat? The answer is, it was created, generated along the metal rod by virtue of the temperature gradient therein. Remember that entropy is a measure of uncertainty, of loss of information, of "mixed upedness" of the energy distribution within a system. This is not hard to create, people do it every day.

When entropy is created, ability to perform useful work is lost. The energy is still there, but it has become degraded. Today we hear much talk about conserving energy. Now we have learned energy is always conserved, according to the First Law. What we really wish to conserve is the availability of energy to perform useful benefits for mankind. This means to avoid the creation of entropy.

Every real process involving motion has some friction, and every transfer of heat at a finite rate requires a temperature gradient. Both create entropy. The more we do to minimize entropy generation the more expensive becomes the equipment required. What energy conservation really means then is, to use less energy, and when you do use it, do so as efficiently as possible, with the constraints of rational economics.

REFERENCES

1. Reynolds, William C. *Energy From Nature to Man,* (Hightstown, New Jersey: McGraw-Hill Book Co., 1974), p. 5.

2. Tribus, Myron. *Thermostatics and Thermodynamics*, (New York, New York: Van Nostrand Reinhold Co., 1961), pp. 1–10.
3. Tribus, Myron. *Thermostatics and Thermodynamics* (New York, New York: Van Nostrand Reinhold Co., 1961), pp. 56–64.
4. Reynolds, William C. *Thermodynamics*, 2nd ed. (Hightstown, New Jersey: McGraw-Hill Book Co., 1968) pp. 137–149.

PROBLEMS

1. a) State your weight in pounds. Determine your weight in newtons and in dynes, and your mass in slugs and kilograms.
   b) State your height in feet and inches. Determine your height in meters. In track meets a race of 1500 m is often called the "metric mile." If a runner has been clocked at 4 minutes flat for a statute mile (5280 ft), estimate his time to run the metric mile.
   c) The acceleration of gravity on the surface of the moon is 5.32 ft/sec^2. An astronaut weighs 200 lb on earth when fully dressed and equipped with his "moon suit" and "life pack." What will be his mass in kilograms and his weight in pounds on the moon?

2. A 4000-lb Buick is at the top of an incline 100 ft in vertical height above a level roadway below. What maximum speed in miles per hour is possible for the Buick at the bottom of the incline if it coasts (without engine power)? What speed would be possible for a 2000-lb Volkswagen coasting down the same incline? How much work is required to get the Buick back to the top of the incline? How much for the Volkswagen?

3. In a mileage economy test of an automobile by the EPA, the vehicle is stationary while the rear wheels turn on a large drum. The torque on the drum is controlled to simulate road resistance, hill climb, etc. This is called a dynamometer test. In simulating a highway mileage test, the dynamometer is set for level road conditions. Does the test automobile perform any work? A one-hour test is performed at a simulated 50 mph and the vehicle consumes exactly 2 gal of gasoline, with an energy content of 130,000 Btu/gal. The dynamometer registers a constant 40 million foot pounds torque during the test. What is the efficiency of the vehicle, *i.e.*, ratio of useful work performed to energy input. How many horsepower are delivered to the rear wheels? (550 ft-lbf/sec = 1 hp).

4. Many inventors have proposed perpetual motion machines. One clever idea has been to stretch a belt over two pulleys and attach to the belt, a series of equally spaced small cylinders filled with air and closed-off by a weighted piston. The device is immersed in water with one pulley directly above the other. On one side, the

weighted pistons act downward reducing the volume of the air in the cylinders. On the other side, the pistons are held from falling out by stops at the end of the cylinder. With the pistons in this position the air volume in the cylinder is larger. As everyone knows there is a buoyancy effect, an upward force when water is displaced by a body of lower density. In this case the buoyancy effect on the cylinders with larger volumes of air exceeds that of the cylinders with compressed air resulting in a net unbalance force, higher on one side of the belt than the other. What do the Laws of Thermodynamics have to say about this device? Will it work? Will the belt move at all? Ask your instructor to explain its motion or non-motion. If he can't maybe you should build one.

5. A common method of increasing the effectiveness of heat exchangers is to add extended surface, fins, to one or both sides. You can observe this by looking at the radiator of an automobile where aluminum plate fins are attached to tubing. Water from the engine block is pumped through the tubes, and air is drawn over the tubes and fins by a fan attached to the engine shaft. The water is cooled by the air and returned to the engine to cool the cylinder walls.

   It is proposed to increase the output of a home heating furnace by attaching nose fins. The entrepreneur shows you two metal boxes each containing a 100-watt light bulb. One is plain, the other has thousands of nails soldered to it, to increase the surface area and presumably to increase the heat output. Will the nails increase the heat output? What other changes will they accomplish, if any, compared to the plain metal box?

6. For fire fighting it is proposed to use a compressed air device to propel a light line to the top of a building. Compressed air at a gage pressure of 690,000 newtons/m^2 is available from a tank on the fire truck. A weight of 2 kg is to be launched to the roof of a building, the weight of the line it carries can be neglected. The 2-kg weight sits on top of a ½-kg piston in a cylinder. At launch the compressed air is introduced behind the piston and acts on it at constant pressure over a distance of 0.5 m in the launch tube. The piston has a diameter of 10 cm. How high can the weight be fired. The atmospheric pressure is 101,000 N/m^2.

7. Someone has paraphrased the First Law of Thermodynamics as, "You can't get something for nothing," and the Second Law as, "You can't get as much as you thought you could."
   Another person has suggested, "You can't win", and "You can't even break even."
   Do these strike you as being appropriate phrases in the context of the First and Second Laws? Explain. Do you have any alternative paraphrases you would like to suggest?

8. An ideal heat engine operates on a Carnot cycle which is illustrated below on a pressure-volume (P-V), and temperature-entropy (T-S) diagram.

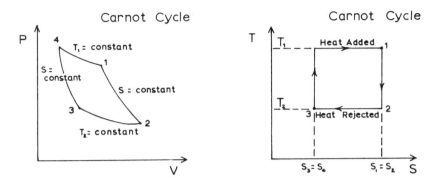

Note that all of the heat is added at a uniform temperature, T_1, and all the heat rejected is from a uniform temperature T_2. The processes 1-2 and 3-4 are called "isentropic" or "reversible adiabatic." Explain the meaning of these two terms. Consider a Carnot Engine that operates between temperatures of 500°F and 70°F. If the desired output of the engine is 100 horsepower, compute

a) the heat supplied in Btu/hr
b) the heat rejected in Btu/hr
c) the engine efficiency
d) the entropy change of the 500°F reservoir supplying the heat in Btu/°R-hr.

9. a) A storage battery (automobile type) can produce work in the form of an electric charge moving through a voltage difference while exchanging heat only with a uniform temperature atmosphere. Is this a violation of the Kelvin-Planck statement of the Second Law? Explain.

   b) Consider a *cyclic* process for a storage battery as follows: Process 1: The battery is "charged" with 2800 watt-hours of electrical work, while 700 Btu flow *out* to the atmosphere. Process 2: 2400 watt-hours of work are drawn from the battery. How much heat was transferred in process 2? (when the battery returned to the same state as before it was "charged" in process 1).
   After process 1 occurred as described above, does the First Law or Second Law limit the maximum possible work of process 2? How much work is possible?

10. A residential "heat pump," as illustrated in the diagram below, extracts heat Q_E from the outdoors at temperature T_E, and de-

livers heat Q_H in the house where the temperature is T_H. The compressor circulates a working fluid called "Freon,®" and is driven by an electric motor, which draws work energy from the power line. The valve reduces the pressure of the Freon and, thereby, the temperature. The entire process experienced by the working fluid is cyclic.

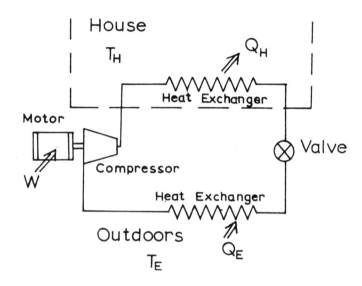

a) What is the relationship between W, Q_H and Q_C in one cycle?
b) What advantage could be claimed for this system of heating a house over directly using electrical energy in resistive heating elements (such as are used in an electric range for cooking)?
c) The valve permits an unrestrained expansion of the Freon from a higher to a lower pressure. This takes place rapidly and in a small area so there is no heat transfer (and no work). Will the entropy of the Freon increase, decrease, or remain the same when it flows through the valve?
d) Suppose the heat pump were reversible (though it is not); that all the heat transferred from the outdoors to the Freon was at a uniform temperature of 40°F, and that all heat transferred into the house was at a uniform temperature of 80°F. For each watt-hour (3.4 Btu) of electrical energy used by the motor/compressor how many Btu of heat would be delivered to the house?

Chapter 4

Electric Power Generation

INTRODUCTION

The basic power plant cycle was discussed in Chapter 3 and illustrated diagrammatically in Figure 31. As was pointed out, the highest thermal efficiency is obtained when heat is added at the highest possible temperature and rejected at the lowest possible temperature. Your knowledge of the basic laws of thermodynamics and of the primary elements of a power plant cycle can now be applied to gaining an understanding of the function and operation of a modern power plant. A more detailed examination of each of these basic elements and some additional ones will be required to do this.

FOSSIL FUEL POWER PLANTS

The steam generator will be examined first. Water from the cycle enters the steam generator and is converted to steam through the addition of heat. Before examining the specific process, it is appropriate to consider the source of heat. The heat may be provided either by combustion of a fossil fuel or by the effects of a nuclear reaction. Nuclear heat sources and reactors will be considered separately in Chapter 5.

Combustion fuels may be coal, oil, or natural gas. Coal predominates and is destined to continue to be predominant over oil or natural gas because both are becoming increasingly scarce and costly. Coal for a power plant is generally obtained from a mine using an entire train operating directly between the mine and the plant. It is stored adjacent to the plant to provide the plant with a fuel reserve of from several weeks to a few months. When needed, it is conveyed to the plant, usually pulverized finer than flour, and finally blown into the furnace for burning. The pulverized coal enters the furnace, which is the main interior of the steam generator, mixed with rapidly moving hot air. The combustion and the effects of the high inlet velocities at the burners cause a large amount of turbulence in the com-

71

bustion area. This helps to insure that each particle of coal contacts oxygen from the air with which to react. The gases created by the combustion of the coal are at temperatures in excess of 3000°F in the combustion region. These gases radiate heat to the furnace walls at high rates as they pass through the interior of the steam generator. The gases are cooled as they transfer heat to the furnace walls and to a variety of heat exchangers located along the path of gas flow. Finally the gases are passed up the exhaust stack and into the atmosphere. It might be noted that, at the point of discharge into the atmosphere, the gases are usually still at a temperature of around 500°F.

When pulverized coal is burned, extremely small particles of fuel enter the combustion chamber. The major part of the particulate mass which remains after the coal has burned is the ash. Ash may typically represent about 10% of the mass of the coal so still smaller particles are formed in the combustion process. Such small particles can be removed from the combustion gases most effectively through the use of electrostatic precipitators. This is usually accomplished just before the gases enter the stack. In many applications over 95% of the ash in the gases is removed and thereby prevented from being dispersed into the atmosphere.

The furnace walls of a steam generator consist largely of pipes containing the water to be heated in the cycle. The walls of a steam generator appear typically as shown in Figure 34.[1] The tubes are vertical and flow is upward. At the bottom, these tubes, which are called risers, contain only water, while at the top a large fraction is steam. The steam is formed because of the large amount of heat transferred. Each riser ends at a device known as the steam drum. The steam drum, in simplest form, is a horizontal header tank in which the separation of steam and liquid takes place. Controls maintain the water level at approximately the midpoint of the drum. The fraction of the flow from the risers which was steam leaves through a pipe at the top of the drum, and the fraction of the flow from the risers which was liquid flows through relatively large pipes down to the level of the inlet to the risers. There it enters a header which distributes it to the risers. These pipes which carry the water downward are located away from the furnace heat. Since they are unheated, the water density in them is constant and greater than the average density of the fluid in the riser. This causes an upward flow in the riser. It is sometimes necessary to augment this natural circulation by means of pumps which force the flow. This is particularly true with pressures over 2400 psi because the difference between the density of the water and steam in the system becomes relatively small at these pressures. A steam generator with circulating pumps of this type is said to have a forced-circulation boiler.

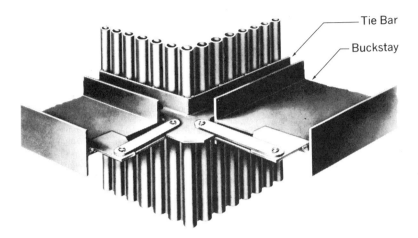

Tie Bar

Buckstay

Figure 34. Risers forming the walls of the furnace section of the steam generating unit. The vertical tubes form a continuous wall. A corner with supporting structure is shown. (Courtesy of Babcock and Wilcox Co.)

After the steam leaves the steam drum it enters the superheater, a heat exchanger consisting of banks of tubes. The steam is heated in the superheater to the highest temperature of the cycle. This is typically in the range of 1000°F–1150°F. A large part of the heat transfer surface of the superheater must be located in the steam generator at a place where extremely high gas temperatures and corresponding high heat transfer rates are avoided. This is necessary to prevent the tubes from reaching a temperature at which they might fail. The problem of tubes overheating is particularly severe in the superheater not only because of the high temperature of the steam inside the tubes but also because the steam inside them does not permit heat transfer as readily as, for instance, the mixture of steam and water in the risers. The tubes of the superheater are made from steel alloys that have high temperature strength and are therefore relatively costly.

A steam-generating unit of the type described is shown in Figure 35. The inlet water flows first to the economizer, which is a heat exchanger. In it, the water is heated to nearly the water temperature in the steam drum, which it enters after leaving the economizer. The economizer conserves energy by using combustion gases after they have given up most of their available heat energy in other heat exchange processes.

Another heat exchanger shown in the figure is the reheat superheater. This device receives steam that has been partially expanded in the turbine. It adds heat to it, usually to bring it to approximately the same temperature as the steam that initially enters the turbine. The reheated steam then flows back to the turbine and reenters it to continue its expansion. The heat added to the steam in the reheater is added at a high temperature. This aids in maintaining high cycle efficiency. Another effect of reheating is to increase the amount of work done by each pound of steam passing through the cycle.

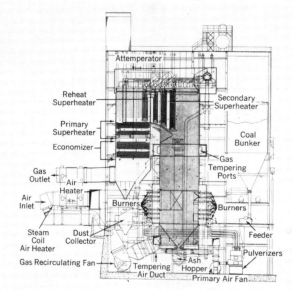

Figure 35. Steam-generating unit for pulverized coal firing. The unit capability is 1,750,000 lb/hr of steam at 1000°F and 2875 psi. The steam generating unit is on the order of 150 ft high. (Courtesy of Babcock and Wilcox Co.).

Finally, note must be taken of the air heater. This heat exchanger uses the heat energy of the combustion gases after they have left the economizer to preheat the air entering from the atmosphere. The combustion gases go to the exhaust stack after leaving the air heater and the heated air becomes the inlet air to the furnace. The air heater serves to reduce the temperature of the gases being rejected to the atmosphere and therefore to reduce the amount of heat rejected in this way. The heat saved serves to reduce the amount of fuel needed to raise the temperature of the inlet air to the desired combustion temperature in the furnace.

The hot combustion gases in the stack are less dense than the colder air in the surrounding atmosphere. For this reason, these gases tend to rise and create a draft. This is called natural draft. The natural draft produced in this manner is generally not sufficient for modern power plants so it is augmented with fans. A fan used to blow air into a furnace is called a forced-draft fan. A fan used to pull gases out of the furnace is called an induced-draft fan. Both are generally required in larger units. The flow through each of these fans is adjusted so that the combustion region of the furnace is maintained at a pressure very slightly below atmospheric. This is done so that any leaks will result in an air inflow and not a loss of hot combustion gases.

A major problem in the operation of the steam generator is water purity. Water in the liquid form is an excellent solvent and in naturally available forms contains quantities of dissolved solids. On the contrary, all solids are virtually completely insoluble in steam. Therefore, dissolved solids precipitate out of solution and are deposited or remain in solution in the liquid fraction when liquid water is converted to steam. The steam leaving the steam drum is essentially pure, and only small moisture droplets which are entrained in the steam flow carry appreciable dissolved solids. All the rest of the dissolved solids remain in the steam drum and are circulated in the boiler water. These impurities can cause scale and corrosion in the boiler tubes, so measures must be taken to continuously remove water containing high levels of impurity and replace it with water as pure as practicable. In steam generating units which operate at pressures exceeding 3206 psi, the water is at a greater than critical pressure. This means that no change of phase from liquid to vapor takes place in the heating process and consequently no steam drum can be utilized. Therefore, a supercritical pressure steam generating unit requires extremely pure water to prevent scale buildup from deposition of solids in critical sections of the tubing. The makeup water in these systems is treated to reduce dissolved solids to far less than one part per million.

The cycle steam flows to the turbine after it leaves the superheater of the steam generating unit. The purpose of the steam turbine is to utilize the energy in the steam to drive an electric generator which in turn generates electrical energy with an alternating current. To produce common 60-cycle current, the generator shaft must rotate at precisely 1800 rpm for a 4-pole or 3600 rpm for a 2-pole generator. The steam turbine must also rotate at this speed since it is connected directly to the generator. This connection may transmit over a million horsepower in some large units. A modern steam turbine and electric generator is shown in Figure 36.

Valves at the steam turbine inlet automatically admit the flow of steam needed to provide the power at the generator. The power output of the total system is controlled at this point. The turbine consists of two principal parts, a casing fixed to the foundation and a rotor which turns on bearings and is directly connected to the generator. The valves, inlet nozzles and the fixed blades are attached

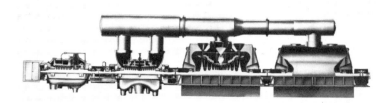

Figure 36. A large modern steam turbine having reheat and four parallel flow paths at exhaust. The unit capability is about 1,000,000 hp. (Courtesy of General Electric Co.).

to the casing. Alternate rows of blades, sometimes called turbine wheels, are attached to the rotor. In terms of steam flow, the inlet nozzles are followed by a moving row of blades or a turbine wheel. The combination makes up the first turbine stage. Large steam turbines have many stages, consisting first of blades fixed to the casing and acting as nozzles, followed by moving blades attached to the rotor. The steam passes through all the stages in succession, flowing parallel to the axis of the turbine rotor. The first two stages of a turbine are shown diagrammatically in Figure 37.

At the point of entry into the turbine, the steam is at a high pressure, temperature and density. Because of the high density, the flow can be accommodated in relatively small passages. The steam expands, however, as it flows through each nozzle on its path through the turbine. A nozzle acts to increase the velocity and therefore, the kinetic energy of the steam. The increase in kinetic energy causes a decrease in the other energy forms of the steam. This is observed as a decrease in its pressure and temperature. The steam flows through the first row of nozzles which are located around the periphery of the turbine casing. These nozzles direct the flow into the first row of blades as suggested in Figure 37. The blades are moving due to the rotation of the turbine shaft and are shaped so that the flow enters smoothly and is redirected as indicated in Figure 37 with minimum friction. The flow is redirected so that the tangential velocity at exit is opposite to the velocity of the blades. This reduces the absolute velocity of the steam and hence its kinetic energy. The kinetic energy decrease is converted into mechanical energy at the rotating shaft. Another way of thinking about this process is to consider that a force is produced by the steam impinging on the moving blades. This force is in the direction of blade movement and produces power as determined through the relationship of power equals force times velocity in the direction of the force.

Understanding the steam flow through the moving blades in large

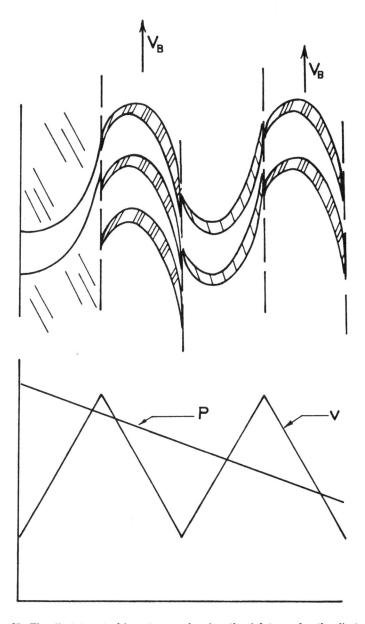

Figure 37. The first two turbine stages, showing the inlet nozzle, the first row of moving blades, the fixed blades, and the second row of moving blades. The variation of pressure P and absolute velocity V as the steam flows through these stages is shown.

turbines is complicated by the fact that these blades also act as nozzles. This means that the steam velocity *relative* to the moving blade increases even though the absolute velocity decreases. Work is done by the reaction force produced by this velocity increase. Turbines with blading as just described are termed reaction turbines since part of the power is produced by reaction. The simple vector diagram of Figure 38 shows the change in the tangential component of the velocity of the steam passing through a moving blade row. There is also a small axial velocity.

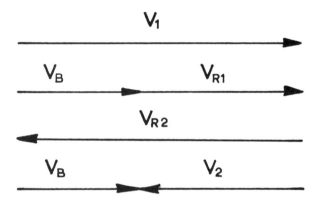

Figure 38. Diagram of tangential velocity vectors for the steam flow in one stage. Absolute velocity leaving the fixed nozzle or blades V_1, blade velocity V_B, velocity relative to the moving blade at blade inlet V_{R1}, velocity relative to he moving blade at blade exit V_{R2}, and absolute velocity leaving the moving blade V_2. V_2 is the velocity entering the next row of fixed blades.

The power produced by each row of moving blades or blade wheels can be calculated if the total mass rate of flow of steam and the steam velocities through the blades are known. Use is made of the relationship

$$F = \frac{ma}{g_c} = \frac{m}{g_c}\frac{\Delta V}{\Delta t} = \frac{\dot{m}}{g_c}\Delta V$$

where ΔV refers to the vector change in the absolute tangential velocity of the steam, and the relationship.

$$P = F\,V_B = \frac{\dot{m}\,V_B\,\Delta V}{g_c}$$

where V_B refers to the blade velocity. For example, let the steam flow be 1,000,000 lb/hr, the blade speed be 400 ft/sec, the entering tangential component of the steam velocity be 1000 ft/sec in the direction of the bláde velocity, and the leaving tangential component of the steam velocity be 600 ft/sec in the direction opposite to the blade velocity. The calculation is as follows:

$$P = \frac{1,000,000 \text{ lb}_m/\text{hr}}{32.2 \text{ ft lb}_m/\text{lb}_f\text{sec}^2} \text{ x } \frac{(1000 + 600) \text{ ft/sec}}{3600 \text{ sec/hr}} \text{ x } 400 \text{ ft/sec}$$

$$P = 5,521,000 \frac{\text{ft lb}_f}{\text{sec}} = 10,000 \text{ hp} = 7488 \text{ kw}$$

The steam expands in passing through each turbine stage so a progressively larger flow path must be provided. This can be accomplished by making the blades longer, but this method, by itself, is limited by blade strength and other problems. When this happens, the steam is withdrawn from the turbine and readmitted at a location along the shaft at which it can be split in two to flow in opposite directions parallel to the turbine shaft. In some large turbines, it is necessary to repeat this process by withdrawing steam from each of these double flow paths and readmitting it at two additional locations along the shaft. At each point of readmission a double flow path is formed, making four parallel paths of steam flow. This is illustrated diagrammatically in Figure 39. Even with four parallel flow paths, blade length in the final stage often exceeds three feet because of the tremendous volume rates of flow. The steam at the exit of the final stage of the turbine is normally at a pressure of less than 1 psig and at a temperature of less than 100°F.

Steam is extracted from the turbine ahead of the exit for reasons other than creating multiple-flow paths. For example, steam is extracted after it is partially expanded and returned to the steam generating unit for reheat. Often extraction for reheat is carried out

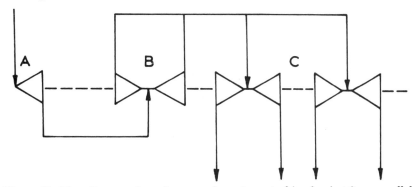

Figure 39. Flow diagram for a large modern steam turbine having four parallel flow paths at exit. Steam enters the single-flow high-pressure section A, is extracted and readmitted in the double-flow intermediate-pressure section B, and is extracted and readmitted in the two double-flow low-pressure sections C.

twice or even three times. Each time the steam is extracted it is at a lower pressure, but it generally is reheated to approximately the same temperature as at the turbine inlet. One of the advantages of reheat is that it reduces the amount of moisture or water droplets formed by the steam as it expands in the low-pressure stages of the turbine. This reduces the amount of erosion damage to the blades. Steam is also extracted from the turbine at several locations for use in feedwater heating, a topic that will be touched upon shortly.

Steam flows from the turbine exhaust at the low pressure of the cycle and enters the condenser which is located directly below the turbine and as near to it as possible. The condenser is a shell and tube heat exchanger in which heat is rejected from the condensing steam and transferred to the surroundings through water circulating through the condenser. The cooling water flows in the ends of the condenser and on the inside of the tubes connecting these "water boxes" on each end of the condenser. The steam flows over the tubes and inside the shell of the heat exchanger. The vapor is condensed on the tube surfaces, falls, and is collected in the bottom of the condenser. This steam, condensed to water, is called condensate. The heat given up in condensing the steam to water is the heat rejection process of the thermodynamic cycle.

The cooling water inside the tubes is usually taken from a river and returned to it several degrees warmer. Considerable quantities are required because about 50 lb of cooling water are needed for each pound of steam condensed. When rivers which are suitable for once-through cooling water supply are not available, especially constructed lakes and channeling may be used. Cooling towers, which allow recirculation of the water, are an expensive last resort.

The condensate collected in the condenser flows down and into condensate pumps, which increase its pressure to somewhat above atmospheric. The water flow later passes through the boiler feed pumps which raise the pressure to that of the high-pressure side of the cycle. Between exiting from the condensate pump and entering the steam generating unit, the water passes through a number of heat exchangers called feedwater heaters. These heaters are usually of the shell and tube type. The feedwater passes through the tubes and steam enters the shell to provide the heat. The steam used is extracted from the turbine. The net effect of the heaters is that the water entering the steam generating unit is heated several hundred degrees above the low temperature of the cycle. Therefore, the heat added from an external source in the steam generating unit is added to the cycle fluid at a higher temperature. This helps obtain a high-cycle efficiency.

The preceding discussion can perhaps be more meaningful through an illustration, using a particular modern power plant.[2] The Marshall Steam Station of the Duke Power Company has been selected for this purpose. This station has four units with a total electrical output capability of 2,136,000 kw. We will focus our attention on only one unit in this station, known as Marshall 3, and will examine conditions of this unit at full capability operation.

The steam generating unit of Marshall 3 is of the supercritical type. The furnace is over 73 ft wide, 36 ft deep, and 158 ft high, the height of a 15-story building. It has 48 burners through which pulverized coal is fed at about 200 tons/hr. The contents of a railroad car carrying 100,000 lb of coal is burned every 15 minutes by this one unit. The flow of water enters the steam generating unit at about 540°F at a rate of about 4,200,000 lb/hr. It leaves as steam at 1000°F and 3500 psi and enters the turbine. It expands in the turbine to a pressure of 927 psi, then is extracted, returned to the steam generating unit for reheat, and finally returned to the turbine at 1000°F. It expands from this condition to 320 psi and the reheating process is repeated, with the steam returned a second time at 1000°F. The steam then expands to the exhaust pressure of about 1 psi. At exhaust, the working length of the turbine blades is over 33 inches. The shaft turns at 3600 rpm and the steam flow is as illustrated in Figure 39. The steam expands through 20 successive stages in the turbine which is direct-connected to the electric generator. The unit capability is 682,000 kw of electrical power output.

After leaving the turbine, the steam flows to a condenser which has 17,032 stainless steel tubes of 1-inch outside diameter. The cooling water for the condenser comes from a 33,000-acre company owned lake. It typically enters at 55°F and leaves at 78°F, flowing at over 300,000 gpm. The condensate passes through the condensate pump, five heaters, and through the boiler feed pump. The boiler feed pumps raise the pressure by over 3500 psi. Two are used, each operating at 5850 rpm and requiring 16,600 hp at full load. The water then passes through two additional heaters before it enters the steam generator.

Up to 20,000 ft³/sec of air is needed to burn the coal. The products of combustion flow through electrostatic precipitators which remove 99% of the particulate matter in the combustion gases before they flow into stacks which are 280-ft tall. It should be noted that each pound of coal burned produces about 3 lb of carbon dioxide in the atmosphere.

Marshall 3 was selected because it exemplifies the best in modern steam power plant practice. Cycle analysis indicates that at the most efficient operating condition, a thermal efficiency of 45.8% can be achieved. In 1969, its first year of operation, a Federal Power Com-

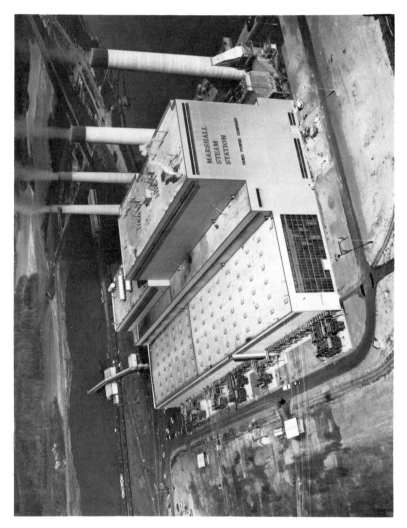

Figure 40. The Marshall Steam Station of the Duke Power Company. (Courtesy of Duke Power Co.)

mission report listed it as, in effect, having the highest thermal efficiency of any steam-electric power plant in the United States. This distinction prevailed for several subsequent years.

Conventional power plants may use coal as fuel or, alternatively, oil or natural gas. Some plants, which were built to use coal as a primary fuel, have equipment installed to allow use of one of these alternate fuels whenever it can be obtained more cheaply, on a dollars per Btu basis, than coal. These alternate fuels have been increasing in cost more rapidly than coal, so usage of this type is now largely curtailed.

Many power plants have been built specifically for burning oil or natural gas only. Such plants can be constructed more cheaply since equipment for handling and pulverizing the coal, cleaning furnace walls, removing particulate matter from the gases going to the stack and disposal of ash is not needed. Plants of this type have been built where oil and natural gas were plentiful and inexpensive, where coal was expensive due to the cost of long distance transportation, or where local environmental legislation prohibited use of coal-fired units. The sharp cost increases for oil and natural gas fuels will be reflected in especially rapid electric rate increases in these areas.

NUCLEAR POWER PLANT

Nuclear energy and the nuclear power plant are discussed in detail in Chapter 5, so only overall comparison of nuclear and fossil fuel-fired power plants is considered here.

The cycle elements are essentially the same for the nuclear and fossil fuel-fired plant with the exception of the steam-generating unit. The conventional steam-generating unit containing the furnace, is replaced with a nuclear reactor in a nuclear power plant. The steam conditions leaving nuclear reactors of current design are far less favorable for a thermodynamically efficient cycle than those leaving the steam-generating unit of conventional plants. Typically, maximum cycle temperatures are on the order of 600°F and cycle efficiencies of around 33% prevail. These lower efficiencies are not involved with inherent operational limitations of reactors, but rather reflect a relatively new technology impeded in its development by massive governmental regulation. Much potential lies ahead for improved nuclear power plant cycles, but delays in the name of safety and environmental concerns may be expected before nuclear power assumes its destined role as a primary source of electrical power.

GAS TURBINE

Gas turbine power plants are utilized for production of electrical energy under specialized conditions. Before discussion of these applications a brief comparison of the gas turbine power plant to the conventional steam plant is helpful.

Our comparison begins at the burner of the gas turbine plant which is comparable to the steam generating unit of the steam plant. It is the location at which heat is added from an external source. Burners of gas turbine plants receive air at several hundred degrees and five to ten atmospheres pressure. Fuel, in the form of oil or natural gas, is admitted and burned. Coal is not an acceptable fuel for most applications because of gas turbine blade erosion from the ash particles that are formed and because of difficulties with its injection into the burner. The amount of fuel added is adjusted to raise the temperature of the gases leaving the burner to the maximum allowable at the turbine inlet. This varies according to design and service but is generally between 1800 and 2400°F.

These hot gases from the burner enter the inlet of the gas turbine which, of course, performs the same function as the turbine in the steam plant. The gases expand through two or perhaps more stages in the turbine and in so doing turn the shaft and produce work. Gas turbines ordinarily run at speeds which are much too high for electric generators so the output shaft of the turbine is connected through reduction gearing to the electric generator.

When the gases leave the turbine they are exhausted to the atmosphere in what is known as the simple cycle. These gases are still at extremely high temperature so utilization is made of this energy in some power plant arrangements in what is known as a regenerative cycle. Suffice to say at this point that the hot exhaust gases are ducted through a heat exchanger and then released to the atmosphere in the regenerative cycle. The hot exhaust gases released to the atmosphere are replaced by air from the surrounding atmosphere. This replacement of a hot fluid for a cold one can be thought of as a heat rejection process. This process then is the one in which heat is rejected from the cycle. It parallels the process in the condenser of the steam plant.

Air from the surrounding atmosphere enters the compressor of the gas turbine power plant and is compressed to the high pressure of the cycle. This compression process raises the air temperature considerably and requires a large portion of the output work of the turbine. The compressor is generally direct-connected to the turbine shaft and is either axial flow or centrifugal flow. Axial flow compressors require several stages to reach the usual operating pressures at output. The compressor is comparable to the pump in the steam plant but requires a vastly higher proportion of the turbine output power. The pump handles a small volume of incompressible liquid while the compressor must handle large volumes of compressible air.

After leaving the compressor, the air enters the burner in the simple cycle, while in the regenerative cycle, air from the compressor passes through a heat exchanger before going to the burner. The air is heated in the heat exchanger by the hot gases from the turbine exhaust. This hot air then requires less fuel to be heated to the desired temperature in the burner. Therefore, less energy from an external source is used and the cycle efficiency is improved. A drawback of the regenerative heat exchanger that makes it unsuitable for aircraft and some other applications is its size. Simple gas turbine plants are extremely compact, but this compactness is compromised by the addition of a regenerator.

Gas turbine power plants are less costly per installed kilowatt of power than conventional fossil fuel or nuclear power plants. They do not need cooling water for operation, and are relatively transportable in smaller sizes. For these reasons they make excellent units for emergency or for meeting short term peak loads. At locations where natural gas is the cheapest fuel and especially if cooling water is at a premium, the gas turbine plant is competitive with the steam plant. Gas turbines can also be combined with conventional steam plants. When this is done, simple cycle gas turbine units are arranged to exhaust into the furnace of the steam generating unit. This unit can utilize part of the energy in the hot turbine exhaust gases. The steam unit can, of course, be separately fired in addition to the gas turbines. The electrical output of the gas turbine units is combined with the electrical output of the steam unit.

DIESEL ENGINE

Diesel engines are widely used for electric power production at locations where the total amount of power required is not large. They compete well with small conventional steam plants because the cost of construction for diesel plants is less per installed kilowatt of power and because the engines have a higher efficiency. Another advantage of the diesel engine plant at some locations is that it requires a minimal amount of cooling water.

Diesel engines are commercially available with power outputs of up to 10,000 kw. Power plants usually have several engines so that maintenance can be performed on one engine while the load is carried by the others. Diesel engines operate best under close to rated load so engines are added and removed from service to meet load fluctuations in these multi-engine plants. Diesel plant capacity to 50,000 kw may be highly competitive with alternate forms of prime power production. It might be noted that large commercial diesel engines are designed to burn the least expensive petroleum fuels and many are designed to operate on a combination of natural gas and oil. Engines which can use both fuels are called dual-fuel diesels. These engines can operate, as needed, from only a small fraction of oil to 100% oil as fuel.

ELECTRICAL GENERATION TO MEET DEMAND

The electrical energy produced in power plants must be produced at the rate of utilization of the energy by the customers of the electric utility. There is no storage of electrical energy in the system. This means that the load on all the generating equipment combined is constantly fluctuating. The utility can control the portion or amount of load each particular unit produces. Normally the most efficient plants operate at full or near full capacity. The less efficient plants are used to meet peak loads and are more likely to be operated at very light loads. In this way, the utility minimizes the operational cost of electric production.

Many problems are caused by the continual need to vary the output, to start up, close down, and maintain power plant equipment. Many elements in the power plant are critical. For instance, the steam-generating unit would catastrophically fail if the feedwater supply were to become interrupted. To protect against this possibility, two or more pumps are used in parallel. One of these is usually steam turbine-driven to insure operation even during a loss of electrical power in the plant.

SUMMARY

A variety of options are available for the generation of electrical energy. Most of it is generated in fossil fuel-fired steam-electric plants today. All indications are that in the future the nuclear plant will become predominant. No single fuel or type of plant is generally superior in all applications, but a system selected at a particular time and location, is in all likelihood superior considering all factors, to the alternatives. The details of nuclear and hydroelectric power are important aspects of this subject which are treated elsewhere.

REFERENCES

1. Babcock and Wilcox Co. "Steam/Its Generation and Use," 38th ed. (1972).
2. Thomas, E. L., I. W. Henry and R. C. Spencer. "Design, Startup, and Operating Experience of Marshall 3 and 4," American Power Conference, Chicago, Illinois (April, 1971).

PROBLEMS

1. A turbine stage produces 3500 hp while receiving a steam flow of 100 lb/sec. The change in the tangential velocity of the steam in this stage is 1500 ft/sec. Calculate the blade velocity in feet per second.
2. Calculate the average horsepower per stage produced by the Marshall 3 turbine at full capability.
3. Calculate the Carnot efficiency of the Marshall 3 unit.
4. Calculate the amount of heat rejected in Btu per minute by the Marshall 3 unit when it operates at full capability. How many pounds of cooling water per minute are needed if the water enters the condenser at 55°F and leaves at 75°F? (1 Btu raises 1 pound of water 1°F).
5. Estimate the tons of coal saved by a nuclear reactor operating for a year with an average output of 1,000,000 kw.

Chapter 5

Nuclear Power

INTRODUCTION

Conversion of nuclear energy to electrical energy on a commercial basis began a little over a decade ago. In early 1974, the United States had about 13 million kilowatts of nuclear power on-line. As we have seen, this represented less than 1% of the electric generating capacity of our country. By 1980, the amount of nuclear power is expected to jump to 150 million kilowatts. It is expected to double again in five years and, by 1990, we could have about 500 million nuclear kilowatts. This is considerably more than our entire national capacity for stationary electric power generation in 1974.

Instead of producing energy by the familiar combustion processes associated with the use of fossil fuels, nuclear energy is derived from the conversion of relatively small amounts of matter. There are available vast amounts of ores containing these nuclear fuels. The challenge is to recover this material economically and convert it into energy.

We begin our study of nuclear power by examining briefly some of the fundamentals of nuclear physics and then turn to the problem of how engineers obtain useful electrical power by controlling nuclear reactions. Last, but certainly not least, we will examine some of nuclear power's attendant hazards and risks.

ENERGY FROM NUCLEAR FISSION

Atomic and Nuclear Structure

Prior to the discovery of radioactivity, near the end of the 19th century, the atom was believed to be indivisible. Subsequently, experiments by Thomson, Bohr, Rutherford, Planck, Chadwick, and many others led to the knowledge that atoms contain many, more fundamental, particles. The atom is an assemblage of neutrons and protons tightly clustered in a nucleus which is surrounded by electrons in a

variety of orbits. Protons are charged particles having a mass equal
to about 1836 times that of electrons and having a unit positive charge
exactly opposite that of the negatively charged electron (1.6 x 10^{-19}
coulomb). Neutrons have a mass slightly greater than that of a proton
but have no electrical charge.

An atom of a particular element may be represented by its atomic
number, Z, which represents the number of protons contained in its
nucleus. In the nonionized state the atom is electrically neutral and,
consequently, Z electrons surround the nucleus in various energy
levels. Alternatively, the mass of the atom is designated by its mass
number, A, which is essentially equal to the number of protons plus
the number of neutrons contained in the nucleus, N. Thus

$$A = N + Z \qquad (40)$$

Many of the elements in the periodic table exist in several isotopic
forms. Isotopes of a given element differ in the number of neutrons
contained in their nuclei (and consequently in their atomic mass),
but are alike in the number of protons and electrons they contain.
For example, three isotopes of uranium occur naturally and their
relative abundances are listed below (abundance is given in atomic
percent):

$$_{92}U^{234} - 0.006\%$$

$$_{92}U^{235} - 0.714\%$$

$$_{92}U^{238} - 99.28\%$$

The presubscript, which is often dropped from the designation since
it is the same for all isotopes of a given element, is the atomic number, Z. The superscript is the atomic mass, A.

The physical atomic mass unit (u) is equal to one-twelfth of the
mass of a carbon atom, $_6C^{12}$. This unit is equal to 1.660531 x 10^{-24} gm
and the masses of the three atomic particles based on the carbon
system are

$$\text{Proton, P} - 1.007277 \text{ u}$$

$$\text{Neutron, n} - 1.008665 \text{ u}$$

$$\text{Electron, e} - 0.00548593 \text{ u}$$

If the masses of the neutrons and protons contained in the nucleus
of a given atom are added, the total will exceed the experimentally
determined mass of that nucleus. The difference between these two
values is the energy that binds the nucleons (protons and neutrons)
together. The binding energy can be represented by Einstein's
famous equation

$$\Delta E = \Delta mc^2$$

where ΔE is the binding energy, Δm the so-called mass defect, and c is the speed of light. The energy equivalent of one atomic mass unit, u, is

$$E = (1.66 \times 10^{-24} \text{gm}) (3 \times 10^{10} \text{cm/sec})^2 = 14.918 \times 10^{-4} \text{erg}$$

or $E = 14.918 \times 10^{-4} \text{erg}/1.6 \times 10^{-6} \text{erg/MeV} = 931 \text{ MeV}$

The energy is expressed in millions of electron volts (MeV). The electron volt is defined as the kinetic energy acquired by an electron as it is accelerated from rest through a potential difference of one volt.

$$(1.6 \times 10^{-19} \text{ coulombs}) \times (1 \text{ volt}) = 1\text{eV} = 1.6 \times 10^{-19} \text{ joule}$$

Although the electron volt is commonly defined by invoking the electron, it is simply a unit of energy and can be applied to any moving particle or to stored energy.

The stability of a particular isotope is related to the binding energy per nucleon for that isotope. However, the nuclei of many of the heavy elements are unstable (radioactive) and these unstable nuclei decay by emitting three types of radiation: alpha particles, beta particles, and gamma rays. The alpha particle is a helium nucleus, having a double positive charge due to its pair of protons which combine with a pair of neutrons to make the mass approximately 4 u. The beta particle is a negatively charged electron which is expelled by the nucleus, effectively converting a proton to a neutron. Gamma rays are electromagnetic radiation which, when emitted, allow the nucleus to drop to a lower energy state.

The degree of stability of a radioactive element is indicated by its half-life; that is, the time period during which one half of a given sample of the element will transform to another element by emitting either alpha or beta particles. The end product of a natural radioactive decay series is a stable element whose binding energy per nucleon ratio is higher than that of the parent element. The radioactive decay series for $_{92}U^{235}$, shown in Table 7, illustrates the relationships between isotopes and the three types of radiation. In each of the decay processes the mass of the resultant particles is less than that of the parent nucleus. The difference may show up as gamma radiaton or as kinetic energy shared by the daughter nucleus and an emergent particle.

Harnessing Nuclear Energy

Both the fission process, where a heavy nucleus splits into two fragments, and the fusion process, where two light nuclei combine to form a heavier nucleus, tend to produce more stable elements. The feasibility of nuclear power depends on utilizing the resultant energy released by these processes.

Table 7. Natural Radioactivity of U^{235}

Element	Half-Life	Type of Radiation (Energy-MeV)
$_{92}U^{235}$	7.07×10^8 yr	$\alpha(4.52)$, $\gamma(0.09)$
$_{90}Th^{231}$	25.6 hr	β, $\gamma(0.03)$
$_{91}Pa^{231}$	3.4×10^4 yr	$\alpha(5.05)$, $\gamma(0.32)$
$_{89}Ac^{227}$	21.6 yr	$\alpha(5.0)$, $\beta(0.22)$
$_{90}Th^{227}$	18.1 days	$\alpha(6.05)$, γ
$_{88}Ra^{223}$	11.7 days	$\alpha(5.86)$, γ
$_{86}Th^{219}$	3.29 sec	$\alpha(6.82)$, γ
$_{84}Pi^{215}$	1.83×10^{-3} sec	$\alpha(7.36)$
$_{82}Pb^{211}$	36.1 min	$\beta(1.4)$ $\gamma(0.8)$
$_{83}Bi^{211}$	2.16 min	$\alpha(6.62)$, β, $\gamma(0.35)$
$_{84}Po^{211}$	0.5 sec	$\alpha(7.43)$
$_{81}Tl^{207}$	4.76 min	$\beta(1.4)$, γ
$_{82}Pb^{207}$	Stable	

Fission, shown schematically in Figure 41, occurs when a fission-able nucleus captures a neutron. The internal balance between neutrons and protons in the nucleus is upset and it splits into two lighter nuclei.

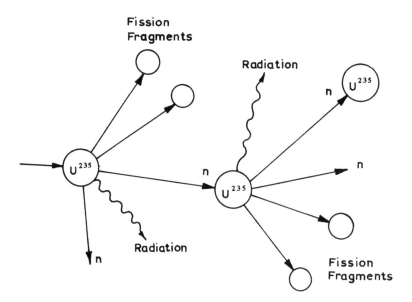

Figure 41. Schematic of U^{235} fission.

During this process an average of 2–3 neutrons is emitted. The mass of the resulting products is less than the sum of the masses of the original nucleus plus the captured neutron and the difference appears as energy according to Einstein's equation. If the released neutrons are captured by other fissionable nuclei more fission events occur. When the reaction becomes self-sustaining so that one fission triggers at least one more fission, the phenomenon is termed a chain reaction. The device in which this chain reaction is initiated, maintained, and controlled is called a nuclear reactor. About 80% of the energy released during a fission nuclear reaction appears as kinetic energy of the fission fragments and neutrons. Through numerous collisions between fission fragments, neutrons, structural material, and fuel, their kinetic energy is reduced and released in the form of heat.

Nuclear Reactor Components

A typical nuclear reactor power plant based on a pressurized-water reactor is shown schematically in Figure 42. Components which are common to most nuclear reactors used as a heat source for power generation are shown in the following table and discussed below.

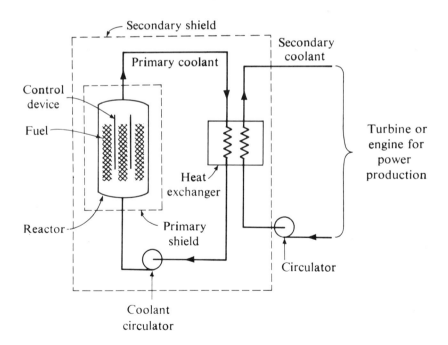

Figure 42. Schematic of a nuclear power plant.[1]

Fuel

A *thermal* nuclear reactor is one in which fission is induced by neutrons in thermal equilibrium with the reactor core material (Table 8). Most of today's reactors are thermal. The nuclear reaction which

Table 8. Reactor Components and Materials.[1]

Component	Material	Function
Fuel	U^{233}, U^{235}, Pu^{239}, Pu^{241}	Fission reactor
Moderator	Light water, heavy water carbon, beryllium	To reduce energy of fast neutrons to thermal neutrons
Coolant	Light water, heavy water, air, CO_2, He, sodium, bismuth, sodium potassium, organic	To carry heat to the heat exchanger
Reflector	Same as moderator	To minimize neutron leakage
Shielding	Concrete, water, steel, lead, polyethylene	To provide protection from radiation
Control rods	Cadmium, boron, hafnium	To control neutron production rate
Structure	Aluminum, steel, zirconium, stainless steel	To provide physical support of reactor structure and components, containment of fuel elements

takes place in these reactors is fuel isotope + neutron → fission fragments + neutrons + energy. For example

$$_{92}U^{235} + _0n' \rightarrow _{37}Rb^{94} + _{55}Cs^{140} + 2_0n' + Q$$

Even though the sum of the atomic masses on both sides of the equation above are equal when expressed as whole numbers, their sums would not be equal if we were to add up the masses of the individual neutrons and protons in the respective nuclei. The difference, the mass defect mentioned previously, appears principally as kinetic energy of the fission fragments. Each fission event yields approximately 200 MeV of energy. This corresponds to 2.3 x 10^4 kwh/gm of material fissioned, and is equivalent to about 3 tons of coal or 14 bbl of crude oil. The thermal energy resulting from the fission process is absorbed by a cooling fluid such as water, helium, or liquid metal, and either transfers the energy to a secondary coolant or is used directly to drive a turbine or an alternator. From that point on a nuclear power plant is similar to a fossil fuel power plant.

U^{235} is just one of several isotopes used in nuclear power generation. Fuel isotopes which are capable of being fissioned by thermal neutrons are classified as *fissiles*. Three additional fissile isotopes are $_{92}U^{233}$, $_{94}Pu^{239}$, and $_{94}Pu^{241}$. These three do not occur naturally but can be made—natural uranium (0.7%), slightly enriched U^{235} (1-2%), $_{90}Th^{232}$, $_{92}U^{238}$, and $_{94}Pu^{240}$. The use of the fertile isotopes will be discussed later.

Of the four fissile isotopes listed above only U^{235} occurs naturally. Furthermore, as mentioned previously, even this fissile isotope constitutes only about 0.7% of uranium metal. Thus, depending on the concentration of U^{235} used in the fuel, a further reactor classification can be made—natural uranium (0.7%), slight enriched U^{235} (1-2%), and highly enriched U^{235} (~90%). Each of these fuels has its advantages and disadvantages. The use of natural uranium eliminates the need for costly separation processes but the reactors are very large and there is a limited group of suitable moderators. Enrichment in U^{235} allows a reduction in the size of the reactor which is dependent on the concentration, but the cost of isotope separation must be considered.

The neutrons produced in the fission of U^{235} have about 2 MeV of energy. Since there are typically 2–3 neutrons produced in each fission event, their energy constitutes only about 5% of the energy produced in the process. The importance of neutrons produced during fission, then, is not their energy but that they determine the possibility of further fission events. Further, since only one neutron is required for fission, the fact that more than one is normally produced makes possible the creation of more fuel than is consumed, through absorption of excess neutrons by fertile isotopes. This phenomenon is the basis for the so-called breeder reaction to be discussed in a later section.

Moderator

Due to the high-energy value of fission neutrons (2 MeV) relative to that required to trigger another fission event (0.025 eV), their probability of interacting with another U^{235} atom is small. The probability of an interaction of a neutron and a bombarded nucleus is referred to as *neutron cross section*. To increase the probability of interaction a moderator is put into the reactor core to slow down the neutrons to thermal energies (a few hundredths of an MeV) by collision with inert atoms. This process is called scattering. A good moderator reduces the energy (speed) of neutrons within a small number of collisions (possesses a high-scattering cross section). Materials consisting of atoms of low mass number usually make the best moderators.

Coolant

The purpose of the reactor coolant is to remove heat released by fission. To do this efficiently the coolant should have a high specific heat, high conductivity, good stability, good pumping characteristics, and low neutron absorption cross section. Coolants can be either liquid or gaseous. Light and heavy water are the principal coolants in use today. Light water is particularly desirable because it is inexpensive and plentiful. Furthermore, it can be used as both coolant and moderator.

Control Rods

Depending on the fate of the neutrons released in fission, the reaction becomes subcritical, critical, or supercritical. For a fission reaction to be self-sustaining, at least one of the fission neutrons must be captured by a fissile nucleus. The term critical refers to a chain reaction in which fission is maintained at a constant rate. Subcritical and supercritical refer to decreasing and increasing fissioning rates, respectively.

Control rods are normally made of cadmium, boron, or hafnium. These metallic elements all have huge neutron absorption cross sections. The control rods are raised or lowered in the reactor core shown in Figure 42. Since the reactor power is directly proportional to the neutron density, lowering the control rods will remove neutrons from the reactor core and will decrease the power and reaction rate. Conversely, raising the control rods will increase the power and rate.

The control rods are a key element in solving the main problem in nuclear engineering, that is, how to harness and control the nuclear chain reaction.

Shielding

Shielding is necessary to prevent passage of radiation to the outside of the reactor. It is provided for two sources of radiation—the neutron and gamma radiation produced in the reactor core, and the gamma radiation present in cooling circuits due to activation of the coolant as it passes through the core. The kind and amount of shielding required are a function of both the type of radiation and its intensity. In general, the best shield for neutrons is a low atomic weight material and for gamma rays a high atomic weight material. Frequently, a reactor shield is constructed in layers of heavy and light materials such as concrete and water. Shields for external circuits, where only gamma radiation is present, may be of various materials. Concrete is most often used because of its low cost.

NUCLEAR POWER PLANTS

Nuclear power plant reactors are generally considered to be of three types: burner reactors, converter reactors, and breeder reactors. These reactors are all based on nuclear fission and may be classified in a number of ways. For our purposes, the energy spectrum of the neutron population, and the use to which the neutrons produced by fission are put, will provide a basis for discussion.

The *thermal* reactor, the principal type in use today, is one in which fission is due mainly to low-energy neutrons. Neutrons produced in fission have an average energy of about 2 MeV each. In a thermal reactor these neutrons are slowed down, that is, their kinetic energy is reduced by collisions with nuclei of the "moderator," typically light water, heavy water, carbon, or beryllium. These moderator elements serve to reduce the neutron energy to the vicinity of 0.025 eV. A *fast* reactor is one that has little or no moderating material and the chain reaction proceeds by way of fission at neutron energies in the MeV region.

Burner and Converter Reactors

The two types of reactors now commercially available in the United States, both of which are thermal or slow-neutron reactors, are the Light-Water Reactor (LWR) and the High Temperature-Gas Cooled Reactor (HTGR). The LWR is a burner reactor; fissile isotopes are consumed but no fertile isotopes are converted to fissiles. The LWR uses ordinary water both as a coolant to transport the heat released in fission and as a moderator to slow down the fast neutrons produced in fission. There are two types of LWRs, the Pressurized-Water Reactor (PWR) and the Boiling-Water Reactor (BWR). The HTGR is a helium-cooled converter reactor which uses graphite as a moderator. A brief discussion of these reactor types, their performance characteristics, and pollutants and hazards associated with their operation is given below. Temperatures, pressures, fuel compositions, and other facts used are typical of the state-of-the-art.

Boiling Water Reactors

Boiling water reactors are generally fueled with uranium dioxide (UO_2) enriched to 2.5–3.2% U^{235}. The fuel pellets are hermetically sealed in zirconium alloy tubes and assemblies of these fuel rods along with control rods are constructed as shown in Figure 43.

A complete reactor assembly is shown in Figure 44.

In a BWR, the primary coolant water is allowed to boil inside the reactor vessel. The steam is separated from the hot water and leaves at about 1000 psi pressure and 546°F. Since the coolant water may

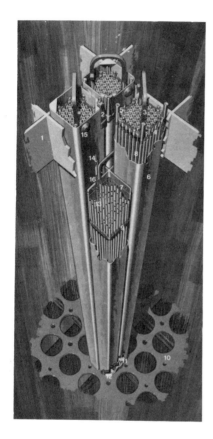

Figure 43. Fuel assemblies and control rod module typical for boiling water reactors. BWR/6 Fuel Assemblies & Control Rod Module: 1. top fuel guide, 2. channel fastener, 3. upper tie plate, 4. expansion spring, 5. locking tab, 6. channel, 7. control rod, 8. fuel rod, 9. spacer, 10. core plate assembly, 11. lower tie plate, 12. fuel support piece, 13. fuel pellets, 14. end plug, 15. channel spacer, 16. plenum spring. (Courtesy of General Electric Company.)

contain impurities which become radioactive upon exposure in the reactor vessel, the steam lines to the steam turbine must be shielded.

Pressurized Water Reactors

A PWR is similar to a BWR except that the primary water circulates only within the reactor vessel. The water reaches about 610°F but the pressure is held at about 2250 psi which prevents the water from boiling. Heat is exchanged between the primary water and secondary water in a heat exchanger contained within the reactor vessel. Since this secondary water doesn't contact the "hot" fuel as-

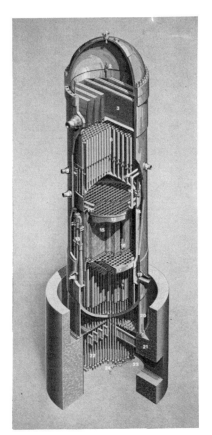

Figure 44. Typical boiling water reactor assembly. BWR/6 Reactor Assembly: 1. vent and heat spray, 2. steam dryer lifting lug, 3. steam dryer assembly, 4. steam outlet, 5. core spray inlet, 6. steam separator assembly, 7. feedwater inlet, 8. feedwater sparger, 9. low pressure coolant injection inlet, 10. core spray line, 11. core spray sparger, 12. top guide, 13. jet pump assembly, 14. core shroud, 15. fuel assemblies, 16. control blade, 17. core plate, 18. jet pump/ recirculation water inlet, 19. recirculation water outlet, 20. vessel support skirt, 21. shield wall, 22. control rod drives, 23. control rod drive hydraulic lines, 24. in core flux monitor. (Courtesy of General Electric Company.)

semblies, the external steam lines in a PWR plant need not be shielded.

The location of principal components in a pressurized-water plant are shown in Figure 45a. Note that the components outside the primary containment building are similar to those found in a fossil fueled power plant. Figure 45b shows a cutaway model of a modern nuclear steam supply system based on a pressurized-water reactor.

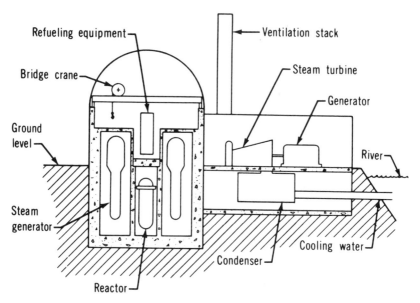

Figure 45a. Location of principal components in a pressurized-water plant. (*Nuclear Power Plants*, USAEC.)

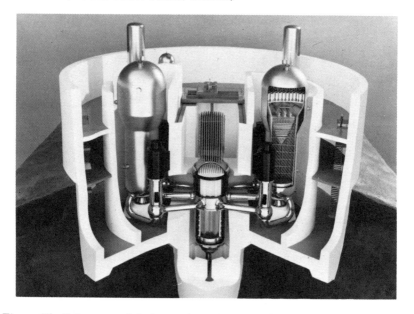

Figure 45b. Cutaway model of a nuclear steam-supply system based on a pressurized-water reactor. Note size of man on stairs at right and operating refueling crane. (Courtesy of Combustion Engineering, Inc.)

High Temperature Gas-Cooled Reactors

The HTGR is classified as a *converter* reactor. It uses a more highly enriched fuel consisting of a mixture of about 5% U^{235}, 95% Th^{232}, and less than 1% U^{238}. The Th^{232} serves as a fertile isotope which is converted to fissionable U^{233} by the reaction

$$_{90}Th^{232} + _{0}n^1 \rightarrow {}_{92}U^{233} + 2 \, _{-1}e^0$$

Fission of the U^{233} takes place subsequent to its conversion from Th^{232}. Thus, the major portion of the energy produced in a HTGR results from the relatively inexpensive thorium. The fuel elements are clad with graphite which serves as a moderator and also permits high temperatures, and, therefore, high thermal efficiency.

Helium, the coolant in many HTGRs, enters the reactor at $-760°F$ and leaves at $-1430°F$ and 700 psia. It flows through a steam generator where steam at 2500 psia and 1000°F is produced, and then returns through a compressor to the reactor. The steam passes through a turbine for power production.

Performance Characteristics

Power output of a reactor is often classified on the basis of fuel burnup. Fuel burnup may be expressed as a) Mw-days per metric ton of heavy metal, Mwd/ton, b) Percent atom burnup, or c) Fissions per cm^3. If all the atoms in a gram of U^{235} could be fissioned, there would be a release of 0.949 Mw-days of thermal energy. This energy release is computed as follows:

$$\frac{6.023 \times 10^{23} \left(\dfrac{\text{atoms } U^{235}}{\text{mole}}\right) \times 200(\text{MeV}/U^{235} \text{ atom})}{235 \left(\dfrac{\text{gm } U^{235}}{\text{mole}}\right)}$$

$$\times \, 4.44 \times 10^{-23} \left(\frac{\text{Mwh}}{\text{MeV}}\right) \times \frac{1}{24} \, (\text{days/hr})$$

$$= 0.949 \, (\text{Mwd/gm } U^{235})$$

If all of the atoms in a ton of fuel were fissioned, nearly a million Mw-days of energy would be produced. Actual burnup figures are over one order of magnitude lower than this due to limitations imposed on the operation by several factors. Buildup of fission products, release of fission gases which contribute to pressure in fuel element cladding, and others, are among the reasons that present-day reactors operate with burnups much lower than the theoretical limit. The following table shows typical values for burnup as well as for specific power.

Table 9. Performance Characteristics[2]

Reactor	Fuel Burnup[a]	Specific Power[b] Current	Predicted
PWR	30,000–35,000 $\dfrac{\text{Mwt-days}}{\text{metric ton U}}$	35 -	46 $\dfrac{\text{kwt}}{\text{kg U}}$
BWR	30,000–35,000	22 -	30
HTGR	100,000 $\dfrac{\text{Mwt-days}}{\text{metric ton (U + Th)}}$		

[a]One thermal megawatt day is equivalent in heat terms to 24,000 kwh, but since the efficiency of generation is only about one-third, one thermal megawatt day results in the production of only about 8000 kwh of electricity.

[b]This is a measure of the power density of the reactor core—somewhat analogous to the horsepower per pound of an automobile engine.

The fuel burnup figures shown in Table 9 must be reduced by about 2/3 if thermal efficiency is to be accounted for. The thermal efficiency (the efficiency with which thermal energy is converted to electrical energy) of PWRs and BWRs is 32–33% in newer plants. This compares with 40% for most modern fossil fuel steam plants and a prevailing average of about 31% for all electric plants in operation. The thermal efficiency of HTGRs is about 40%, much higher than light-water reactors.

Before proceeding to a discussion of breeder reactors, a brief comparison of today's thermal nuclear reactor and fossil fuel power plants is made. A 1000 Mwe fossil-fueled plant consumes over two million tons of fuel per year. A nuclear plant of the same capacity needs only about 35 tons of uranium dioxide. By comparison with reprocessing of 35 tons of spent nuclear fuel, a 1000-Mw coal-burning plant emits around 10 million tons of carbon dioxide per year and several hundred thousand tons of sulfur dioxide, nitrogen oxides, and ash particles.

The cost of electricity produced by either fossil fuel or nuclear fuel depends on many factors. The rates at which these factors change makes such a comparison difficult. However, one recent comparison will be given and the reader is cautioned that, although the figures are typical for 1974, they are obviously subject to change. In a study of nuclear power versus fossil power costs based on an analysis of 21 utilities, the following figures are obtained: Nuclear power costs averaged 10.52 mills/kwh compared to 17.03 mills/kwh for fossil. Of these costs, 2.15 mills/kwh and 11.25 mills/kwh were the respective nuclear and fossil fuel costs. More importantly, while the nuclear fuel costs had risen only negligibly from 2.12 mills/kwh in

1973, fossil fuel costs over the same time period had risen from 6.35 mills/kwh, an increase of 77%.

As larger numbers of light-water reactor plants are built and put in operation, the requirement for nuclear fuels will necessitate processing ores of increasingly lower quality. Thus, the nuclear fuel prices are expected to increase significantly.

Breeder Reactors

Development of commercial breeder reactors and construction of them in substantial numbers by the end of this century is probably one of the key elements in solving our future energy needs. Our present 53 commercial burner reactors and the 182 more plants committed to construction, which generate power by fissioning less than one percent of natural uranium (the U^{235} isotope), will consume the total amount of economically usable nuclear fuel in about 30 years. On the other hand, breeder reactors, when used in concert with the thermal reactors, will multiply our uranium fuel reserves by more than a factor of 2000.

Breeder reactors derive their name from the fact that they produce more fissionable material than they consume. Breeder fuel consists of a mixture of both fertile and fissile materials. The number of neutrons released is sufficient to propagate the fission reaction, even allowing for some neutron losses, and to produce more fissionable material by the conversion of fertile isotopes to fissile isotopes.

The two most common breeding cycles are those a) in which fertile U^{238} is converted to fissionable Pu^{239}, and b) in which fertile Th^{232} is converted to fissionable U^{233}. Fertile U^{238} is both abundant and inexpensive. It constitutes over 99% of naturally occurring uranium. Furthermore, it exists in large stockpiles as a by-product of past uranium fuel-enrichment processing. In fact, estimates indicate that by the year 2000 we will have enough U^{238} on hand to fuel all of the breeder reactors projected for the 21st century.

The uranium cycle uses high energy neutrons and is carried out in a so-called *fast* reactor. These reactors use liquid sodium metal or pressurized helium as a coolant and no moderator to slow down the neutrons. The thorium cycle is similar to the uranium cycle except that it works best in a thermal reactor. This cycle was considered previously in the HTGR converter section.

The three main types of fast breeder reactors being studied in the United States are the Liquid Metal Fast Breeder Reactor (LMFBR), the Gas Cooled Fast Breeder Reactor (GCFBR), and the Molten Salt Breeder Reactor (MSBR). The latter reactor uses the thorium cycle while the former two use the uranium cycle.

The basic feasibility of the breeder reactor as a power producing device has already been proven. Experimental and demonstration facilities exist in the United States, Great Britain, France, and Russia. Additional development is needed to insure that the large breeder reactor of the future will provide dependable and economical power.

Liquid Metal Fast Breeder Reactor

An LMFBR is shown diagrammatically in Figure 46. The fuel consists of 80 w/o UO_2 and 20 w/o PuO_2 in small diameter stainless steel clad tubes operating at temperatures of 1250–1300°F. Fuel assemblies, clusters of fuel pins, are immersed in liquid sodium coolant which flows at low pressure through the reactor, entering at 750°F and leaving at 1150°F. (As in the discussion of light-water reactors, the values of temperature and pressure here are only typical.) Since the primary sodium becomes intensely radioactive in the reactor, a secondary, nonradioactive sodium coolant loop is interposed betweeen the primary sodium loop and the steam generator. Two different designs of the primary system are being considered—a pot type in which all of the components in the primary circuit are submerged in a sodium-filled pot and a loop type system in which the reactor vessel, pump, and heat exchanger are linked by piping through which sodium is circulated.

Demonstration LMFBR plants are based on specific powers of about 1000 kwt/kg fissile (which at an enrichment of approximately

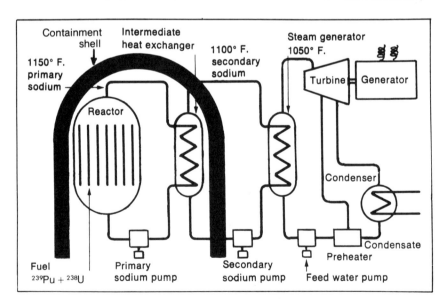

Figure 46. Schematic of a liquid metal fast breeder reactor.[3]

15% fissile corresponds to 150–200 kwt/kg(U + Pu) and fuel burnups of 100,000 Mwt-days/metric ton (U + Pu). The LMFBR is expected to have a doubling time* of six to seven years, a thermal efficiency of about 40%, and a breeding ratio of about 1.5**.

Gas-Cooled Fast Breeder Reactor

The GCFR is cooled with helium gas at 1250 psi and therefore benefits from technology already developed for the HTGR. Unlike sodium, helium does not become radioactive under neutron bombardment, so steam can be generated directly without need for a secondary loop.

The GCFR is expected to have a thermal efficiency of about 40%, a doubling time of 8–9 years, and a breeding ratio of 1.5. The design burnup and specific power are 100,000 in Mwt-days/metric ton (U + Pu) and 800–900 kwt/kg fissile.

Molten Salt Breeder Reactor

MSBR technology draws on the Aircraft Reactor Experiment. Although its technical feasibility has been proved, the present commitment to develop the technology and to construct a molten salt demonstration reactor is not large.

Pollutants and Hazards

Much public attention has been given to the pollutants and hazards associated with nuclear power. Those items most frequently mentioned are radioactive emissions and radioactive wastes, reactor safety and security, and waste heat. Each of these concerns will be discussed.

Before proceeding to such a discussion, however, it is interesting to note how strongly polarized are the technical and nontechnical communities on these issues. R. Philip Hammond, who has had over 30 years experience working with nuclear weapons, reactor fuels, fission wastes, and experimental reactors, writes in the American Scientist,[4]

> "Research in safety and attention to risks have been the watchwords throughout the years; the nuclear industry is without exception the safest in the world in which to be employed, and nuclear hazards are far better understood than those of thousands of widely used chemical and biological agents, or of common energy sources such as coal."

On the other hand, Ralph Nader and Friends of the Earth have recently asked for the shutdown of all operating nuclear power plants.

---

*Time required to double the inventory of fissionable material.
**Mass ratio of fissionable material produced to that consumed.

They state that " the amount of radioactivity routinely present [in one of these plants] is equivalent to 10 times the amount of radioactive fallout from detonation of the largest nuclear weapon in the United States defense arsenal" (complaint filed in U.S. District Court).[4] This statement is an example of the type of reaction faced by the nuclear community. While the statement by itself may be true, no mention is made of the extensive containment facilities surrounding this radioactive material.

An extensive educational campaign will probably be required before much of the public's fears are allayed. As the nuclear power industry grows and its safety record becomes better known, public understanding and acceptance will probably increase.

Reactor Safety

In anticipation of the perennial question, it should be emphasized early in this discussion that there is widespread agreement, even among critics of nuclear power, that an atom bomb type of explosion of a nuclear reactor is absolutely impossible. It is certainly realized, however, that there are hazards associated with nuclear power as with anything else. The objective of nuclear reactor safety has been to reduce this hazard to levels as low as possible consistent with society's needs. This means simply that risks associated with nuclear reactor operation are lower than those society faces every day.

The safety standards applied to nuclear power plants are far more stringent than those applied in any other sector of our society with the possible exception of our manned space program. Since nuclear reactor safety requirements must be based on theoretical potentials, an ultra-ultraconservative approach has been taken.

The principal potential hazard to public safety, presented by a nuclear power plant, is in the large amount of radioactive fission products. Any accident or process that would result in a substantial release of them to the environment could be a threat to public safety.

The basic safety approach in the nuclear industry has led to the concept of "defense in depth." Included in this concept are highly reliable, fail-safe equipment, duplicate and triplicate systems and components, diversity of operating principles for safety equipment, and the design basis accident (DBA) analysis. The multibarrier containment of fission products in light-water reactors, shown in Table 10 and illustrated in Figure 47, is an example of "defense in depth." It would require an extraordinary combination of accidental events to penetrate the multiple barriers and result in any significant release of radiation to the environment.

In an attempt to quantify speculation that arises in reactor safety, the AEC and the nuclear industry have evolved a postulated accident called the design basis accident. Each time the consequences of a

Table 10.

Barrier	Effectiveness
1. Ceramic fuel pellets	Only a fraction of the gaseous and volatile fission products leave the pellets.
2. Metal fuel tubes	Contain fission products which leave the pellets.
3. Reactor vessel and piping	8–10 inch thick vessel—contains reactor cooling water. A small portion of the circulating water is passed through a trap to keep the radioactivity low.
4. Concrete shield	7–10 ft thick—shields operators and equipment from core radiations.
5. Containment shields (primary and secondary)	Encloses the entire reactor part of the plant to prevent or control release of radioactivity in case of reactor cooling water pipe leakage or rupture.

Figure 47. Typical containment for a boiling water reactor. (Courtesy of General Electric Company.)

hypothetical accident are analyzed the plant is redesigned so that there is less threat to public safety. This process is continued, considering events which are more and more improbable, until a point is reached which is agreed by both designers and reviewers that the possibility of such an accident occurring is incredibly small. The extremely unlikely accident just short of that point is termed the DBA. The DBA for light-water reactors is the rupture of one of the large pipes that brings the coolant to the reactor vessel. The consequences of this accident are controlled by the emergency core cooling systems.

Emergency core cooling systems (ECCS) differ depending on the type of reactor. Current PWRs have a number of systems that flood the reactor core with water and prevent its heatup. BWRs have both a system which sprays water on the core from the top and other systems that flood the core. In each case, there are at least duplicate pipe systems and pumps. The probability of several successive failures in parallel systems is negligible.

Just how good is the nuclear power plant record? Since the early days of nuclear power development some 25 years ago, there have been about 2000 reactor-years of operating experience. Many kinds of failures have occurred, most of them trivial, such as broken pumps to be replaced. However, in the United States, no property damage or personal injury to the public or operating personnel has occurred in commercial reactor operations as a consequence of the fissionable or radioactive character of the process.

Radioactive Emissions from Power Plants

Fission products which collect in the UO_2 fuel are the main source of radioactivity. Escape of significant amounts of this radioactivity is prevented by the use of multiple barriers. On occasion, leaks develop in the cladding materials of the fuel rods in PWRs and a small fraction of those fission products which are water-soluble appear in the primary water. The primary water also carries corrosion products made radioactive by neutron bombardment. The radioactive content of the primary water is controlled by continuous purification by filtration and ion exchange. Escape of radioactivity from the primary water is effectively controlled by using a leak-tight pressure vessel and piping system.

The control of radioactivity from BWRs is more difficult because the same water that passes over the fuel elements also passes through the condenser, where it is separated from the cooling water by a metal wall. A rupture of a condenser tube could allow some of the slightly contaminated primary water to mix with the secondary coolant.

Release of radioactivity to the environment from light-water reactors can be controlled to any degree desired, but with increasing cost. All U.S. nuclear power plants are monitored by the U.S. Public Health Service and have been found to add to the environment only a minute fraction of the radioactivity naturally present. For example, the three most prevalent isotopes discharged from a 1000 Mw PWR are H^3, Cs^{134}, and Cs^{137} with respective half-lives of 12.3, 2.3, and 27 years. The ratio of their concentration to the maximum concentration permitted for unrestricted water bodies by the U.S. Public Health Service is only 7×10^{-4}, 2×10^{-7}, and 2×10^{-7} respectively.

The problems with control of radioactive emissions in LMFBRs are considerably more severe than for light-water reactors. The sodium becomes highly radioactive and in addition, the metal sodium is highly reactive with water. Thus, the requirement for leak-tight primary and secondary loops is crucial. However, as in the preceding case, the technology for maintaining emissions at acceptable levels exists, it simply adds to the cost of the electrical power.

The GCFR has an advantage over the previously mentioned reactors in its use of helium as a primary coolant. Helium will not become radioactive and its chemical inertness means there are no cladding-coolant, fuel-coolant, or steam-coolant reactions to design for.

Radioactive Wastes

Spent fuel assemblies from light-water reactors are allowed to "cool" (their radioactivity allowed to decrease) for about four months before reprocessing. After uranium and plutonium are recovered by solvent extraction, the radioactive fission products are concentrated and stored in solution in double-walled containers for about five years. The U.S. Atomic Energy Commission then requires that the solution be evaporated to dryness and the solid fission products, a black, ceramic clinker, be sealed in stainless steel containers. Finally the containers are shipped to a national repository for long-term storage.[5]

With the projected growth of nuclear power to about 9×10^5 Mw by 2000, the radioactive waste disposal problem is not as straightforward as it may appear in the preceding discussion. More recent proposals call for separating the wastes into two categories. The major one in terms of short-term radioactivity is comprised of the fission products, that is, atoms of medium atomic weight. The principal ones are Sr^{90}, Cs^{137}, and Kr^{85}. These isotopes have half-lives not exceeding 30 years, and in 700 years less than one ten-millionth of their activity remains. Seven hundred year storage is considered short-term when one considers storage of the actinides (the elements actinium, thorium, uranium, neptunium, plutonium, and so on) which are formed not by fission but by neutron absorption into the original fuel. All of these are very toxic and have half-lives up to 25,000 years (Pu^{239}).

The decay of radioactive isotopes by the emission of alpha or beta particles is such that the activity (the number of disintegrations/sec) of a given sample decreases exponentially with time. The chance that any particular nucleus will decay at a given instant is independent of the past or future disintegrations of other nuclei. Thus, the overall behavior of these random, independent events can be represented by the equation

$$I = \frac{dN}{dt} = -\lambda N \tag{41}$$

where I is the disintegration rate, N is the number of nuclei present at a given time, and λ is a proportionality constant between them, called the disintegration constant. Separation of the variables and integration between N_0, the number of nuclei present at t = 0, and N, the number present at time, t, gives

$$N = N_0 \exp(-\lambda t) \tag{42}$$

If we differentiate this expression to find the activity or intensity, I, in terms of the original intensity, I_0, we have an exponential relation for radioactive decay:

$$I = \frac{dN}{dt} = -\lambda N_0 \exp(-\lambda t) = I_0 \exp(-\lambda t) \tag{43}$$

Taking the logarithm of Equation (43) gives

$$\ln \frac{I}{I_0} = -\lambda t \tag{44}$$

The half-life, $t_{1/2}$, of radioactive isotopes such as Pu^{239} is the time it takes for the original activity to decrease by one half. Thus

$$\ln \frac{1}{2} = -0.693 = -\lambda t_{1/2} \tag{45}$$

The number of nuclei (or their activity) will thus fall to one-fourth by the end of two half-life periods, and so on. In general, the fraction of an initial sample remaining after *n* half-life periods is $(1/2)^n$.

The *curie* is defined as the quantity of any radioactive species decaying at a rate of 3.7×10^{10} dis/sec. Historically, the curie was based on the estimated activity of one gram of radium. The mass/curie of any radioactive specie may be determined as follows. The rate of decay of N atoms of atomic weight, A, is

$$\text{Rate of decay} = \frac{0.693N}{t_{1/2}} \text{ dis/sec} \tag{46}$$

if $t_{1/2}$ is in seconds. Since there are $N_A = 6.023 \times 10^{23}$ atoms/gm mole, G grams will decay with a rate given by

$$\text{Rate of decay} = \frac{(0.693)\ (6.023 \times 10^{23})G}{At_{1/2}}\ \text{dis/sec} \qquad (47)$$

Thus, the number of curies represented by G grams is

$$\text{Number of curies} = \frac{\dfrac{(0.693)\ (6.023 \times 10^{23})G}{At_{1/2}}\ \text{dis/sec}}{3.7 \times 10^{10}\ \text{dis/sec/curie}}$$

$$= \frac{1.13 \times 10^{13}G}{At_{1/2}} \qquad (48)$$

Finally, the mass of material having an activity of 1 curie is

$$\text{Mass per curie} = \frac{At_{1/2}}{1.13 \times 10^{13}} \qquad (49)$$

Present standards call for extraction of U and Pu from spent nuclear fuel to the extent of 99.5%. Thus, these wastes remain radioactive (toxic) for tens of thousands of years. The technology exists for reducing the elements Ac to Pu to 99.9999% and the less prevalent elements Am to Es by 99%. If we are willing to absorb the cost of these extractions into the price of the electrical power, the final wastes would be reduced to a "nontoxic" level in 1000 years. The extractions would be recycled into the nuclear reactor.

Several locations for long-term, deep underground burial of solid radioactive wastes have been investigated. These are the AEC production facilities at Hanford, Washington, in local basaltic rock, salt beds in Kansas, and salt beds in New Mexico. The long-term safety of such storage depends on preventing the intrusion of water into these locations. The very presence of the salt beds guarantees that no water has been present in the recent geologic past.

The major technical and economic uncertainties involved in long-term storage of radioactive wastes have been resolved. The problem which remains is social; that is, long-term storage means that future problems, if they arise, are passed on to our decendents. Also, the problem of sabotage and terrorism is much in the mind of the public. These questions need intensive study.

Waste Heat

The discussion of waste heat removal in the preceding chapter applies to nuclear power plants as well as fossil fuel power plants. However, since today's LWRs have lower thermal efficiencies, 33 versus 40, more heat has to be removed from a nuclear plant of the same capacity. The temperature rise of the cooling water in a nuclear plant is typically 20°F for a 1000 Mwe, 33% efficient plant and the flow rate is about 1500 ft³/sec.

SUMMARY

Increasing development and utilization of nuclear power throughout the world appears to be inevitable. Hopefully, the highly desirable fossil fuels such as crude oil and natural gas can be conserved and used in applications for which nuclear power is not feasible. There are many detractors to nuclear power generation but the technology exists for producing clean, safe power by nuclear means. The nuclear industry has attained an admirable safety record thus far and we must insist that it continue to do so.

REFERENCES

1. Foster, R. A. and R. L. Wright, Jr. *Basic Nuclear Engineering,* 2nd ed. (Boston, Massachusetts: Allyn and Bacon, Inc., 1973).
2. Schurr, S. H. *Energy Research Needs,* Resources for the Future, Inc., Washington, D.C., distributed by National Technical Information Service, PB 207 516, U.S. Department of Commerce, 1971.
3. Benedict, M. "Electric Power from Nuclear Fission," *Technol. Rev.* (October-November, 1971).
4. Hammond, R. P. "Nuclear Power Risks," *Amer. Scientist* **62**, 155-160 (1974).
5. Kubo, A. S. and D. J. Rose. "Disposal of Nuclear Wastes," *Science* **182**, 1205-1211 (1973).

SUGGESTED FURTHER READING

1. El-Wakil, M. M. *Nuclear Energy Conversion,* (Scranton, Pennsylvania: Intext Educational Publishers, 1971).
2. "Nuclear Power and the Environment," Delaware Valley Section of the American Nuclear Society, P.O. Box 1535, Philadelphia, Pa., 19105.
3. *The U.S. Energy Problem,* Vol. II, Inter-Technology Corp., distributed by NTIS, PB 207 518, U.S. Department of Commerce, 1971.
4. Weinberg, A. M. and R. P. Hammond. "Limits to the Use of Energy," *Amer. Scientist* **58**, 412-418 (1970).

PROBLEMS

1. In 1970, 87 million passenger cars drove 859 billion miles and consumed 63 billion gallons of fuel. Assuming the average speed was 35 mph and each vehicle developed an average of 120 hp, determine the amount of U^{235} in pounds required to supply this energy. Let the overall conversion efficiency be 25%.
2. There are approximately 716 billion barrels of oil reserves known to exist in the world. Determine the energy equivalent in terms of U^{235}.

3. A nuclear reaction such as

$$_{90}Th^{232} + {}_0n^1 \; {}_{92}U^{233} + 2_{-1}e^{\circ}$$

may be written in an abbreviated form $_{90}Th^{232}(n,2e)_{92}U^{233}$. Al^{27} has been made to undergo some 16 nuclear transformations. Write the nuclear equation for each of the following:

(a) (a,n)   (c) (p,a)   (e) (a,2p)
(b) (a,p)   (d) (p,n)   (f) (n,p)

4. Write a possible equation for the fission of U^{235} if one of the fission fragments is Sr^{95} and two neutrons are produced.

5. The half-life of Co^{60} is 5.3 years. A sample of 400 mg of pure Co^{06} is purchased for medical purposes. How many mg will remain after 12 years?

6. (a) Calculate the decay constant λ for U^{235}, whose half-life is 7.07 x 10^8 years. (b) How many disintegrations per second are there in a 1-gm sample of U^{235}? (c) What number of curies in this?

7. Carbon dating of archeological materials of organic origin is based on the fact that the organism, upon death, stops absorbing radioactive carbon 14 as $^{14}CO_2$ from the atmosphere. This radioactive carbon accounts for about 0.10% of the total carbon content. The ratio of C^{14} in a certain wood sample to that in a comparable piece of new wood is measured to be 0.150 $\pm$ 0.001. How old is the wood, and what is the uncertainty in its age? The half-life if C^{14} is 5568 years.

8. A rock returned from the surface of the moon has a ratio of Rb^{87} to Sr^{87} atoms equal to 14.45. Assuming that all of the Sr was Rb upon formation of the solar system, estimate the age of the solar system. The half-life of Rb^{87} is 4.7 x 10^{10} years.

9. Compared to coal costs of $54/ton delivered, what must be the cost of *natural* uranium metal to give equivalent unit energy costs?

Chapter 6

Energy Storage

INTRODUCTION

The ability to store energy when it is not needed, and retrieve it to satisfy a demand is an important requirement for many energy supply systems. In this chapter we are concerned with storage of energy to perform work when it is recalled from storage. In Chapter 9 on Solar Energy the problem of thermal energy storage, or in every-day language, heat storage is addressed.

The storage of energy to perform work is usually accomplished in two ways—by mechanical energy storage or electrical energy storage. Pumping water to a higher level, then returning it through a turbine can be used on a large scale for mechanical energy (potential energy) storage. The "spinning-up" of a flywheel to high rotational velocities is another method of mechanical energy storage (kinetic energy). Compressing air to high pressures and storing it in steel tanks or, on a larger scale, in underground caverns is a third method of mechanical storage. We shall present applications of each of these methods here, as well as the use of the battery for electrical energy storage.

To produce large amounts of work energy, such as electricity, requires large power plants. The capital investment required for such plants is high, and becoming more expensive each year. In 1975 the project cost of construction for conventional steam-electric power plants is in the range of $350–$450 per kilowatt of capacity.

Atlanta, Georgia had a peak demand of about 1800 Mw in 1970, and has a projected demand of near 3500 Mw in 1980. If the difference of 1700 Mw was constructed at an average $400/kw, the capital cost would be $680 million. The objective of large-scale energy storage systems for electric utilities is to reduce such capital investments. Excess electrical energy is generated at night and stored when demand is low. It can then be retrieved, with small losses, during the daytime when demand is high. The cost of the energy storage and retrieval system must, of course, be considerably less than the reduced power plant cost to achieve the desired objective.

In the past, electric utility companies have used gas turbines, or older steam-generating equipment to meet peak power demands. As the costs of petroleum and coal have dramatically increased, the poor economy of this operation has cut into profits. Storage systems look like a more attractive investment, and are being seriously studied by more electric companies.

The development of light-weight high-strength composite materials has brought renewed interest in a very old energy storage device, the flywheel. The prospects for flywheel-driven buses, which were once used on Swiss mountain roads, look promising. In what follows we shall look at the specific requirements for flywheel energy storage.

Lead-acid batteries have been starting our automobiles for years. Pocket calculators and electronic flash guns operate from nickel cadmium batteries, which can be recharged thousands of times. But what are the prospects for electric automobiles? The characteristics of batteries needed for such an application will be discussed in Chapter 7.

POWER DEMAND

The demand for electrical energy varies on a daily and also on a seasonal basis. A number of years ago the electrical demand was highest for the United States during the winter season. The installation of air conditioning in residences and commercial buildings on a large scale in this country has now shifted that peak to summer. In some cities it has been necessary to reduce voltage levels on hot summer days to avoid power failures due to an overload of electrical energy usage. This has added a new word to our vocabulary—"brownouts."

Figure 48 plots the average monthly energy usage in 1965 for the Southern Company System, which serves Georgia, Alabama, and a small part of Florida and Mississippi. The peak summer load of about 4.2 billion kwh occurs in August. The peak winter load is 3.5 billion kwh and occurs in December. Minimum monthly consumption for 1965 occurred in February when it was about 2.9 billion kwh. Figure 49 shows the hourly demand for electrical energy that occurred on the peak summer day of August 6, 1965; and the peak winter day of December 7, 1965. The ratio of power demand at 10 a.m. to that at 3 a.m. for August 6 is 6.85/3.6 or 1.90. The data for these two figures are taken from a report of The Southeast Regional Advisory Committee.[1]

Normally the electric utilities will have about a 20% reserve generating capacity over the peak demand. If this were the case for the Southern Company System, it would have been necessary to have a

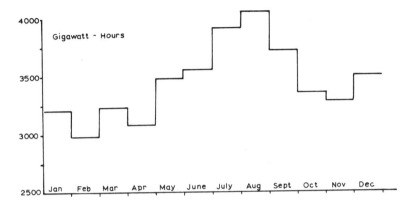

Figure 48. Average monthly electrical energy usage for southeastern region of U.S. in 1965.

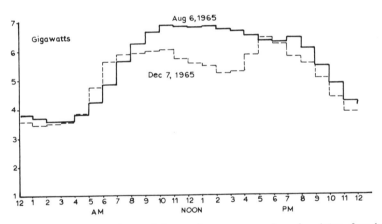

Figure 49. Hourly electric demand for peak summer and peak winter day in 1965, southeastern region of U.S.

generating capacity of 8.2 gigawatts (Gw) in 1965. Total electrical energy usage for the 12 months shown in Figure 48 was 41,300 Gwh. The percentage of available electric power which would have been used can be calculated as

$$\frac{41,300}{8.2 \times 24 \times 365} = 0.575 \text{ or } 57.5\%$$

Because of the seasonal and daily variations in power demand, it is necessary to have nearly twice as much capacity for generating electricity as would be needed for a constant load.

Utility companies meet much of the peaking in power demand by the use of special generating facilities brought on-line for the peak periods. This includes older steam-generating equipment of lower efficiency, gas turbines, and diesel engines. The lower efficiency of older equipment, and the higher per kilowatt cost of smaller capacity equipment, together with their short operating time, adds substantially to the cost of electricity. There is also loss of economy in the startup and shutdown periods and in part-load operation of heat engines. All of the above factors suggest that there may be considerable advantages to the use of an energy storage facility. Excess generation of power during low demand periods, and its return to the distribution system during high demand, serves to produce a more uniform loading of the generating equipment. Let us examine methods of energy storage with respect to their capacity, efficiency, and cost.

MECHANICAL ENERGY STORAGE

Pumped Water Storage

The method of energy storage used almost exclusively by electric utility companies is pumped water storage. The Seneca pumped storage plant[2] near Warren, Pennsylvania, is schematically illustrated in Figure 50. This plant has been in commercial service since January 1970. It has a capacity of 470 Mw.

Unit 1 and unit 2 each have a capacity of 220 Mw. Each has a motor/generator that drives or is driven by a reversible pump/turbine. Unit 1 moves water back and forth between the Allegheny reservoir formed by the Kinzua dam, and a 6000 acre-foot upper Seneca reservoir. Unit 2 can operate either like unit 1, or can discharge to the Allegheny river downstream of the dam. Unit 3 has a capacity of 30 Mw. It is not reversible, and acts only as a generator with stored water flowing from the Seneca reservoir to the river.

There are four major factors to be considered in the development of pumped water storage facilities: a) capital investment, b) availability of large elevated area for water storage, c) environmental effects, and d) overall efficiency of operation.

Obviously the capital investment and efficiency of operation must be such as to make the pumped storage facility economically attractive over additional generating facilities, or other storage methods. The overall efficiency for power generation is the product of the efficiency during pumping and the efficiency during drawdown through the turbine. Typical values for pump-motor and turbine-generator efficiencies are 88% and 85%, respectively. The overall efficiency is then (0.88) x (.85) = 0.75, or 4 kwh of energy from the system during off-peak, returns 3 kwh during peak demand. The

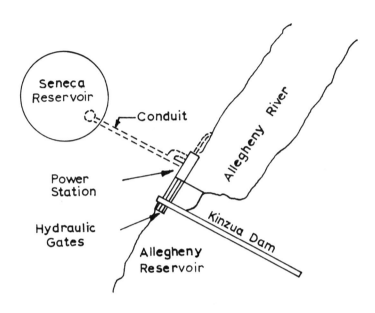

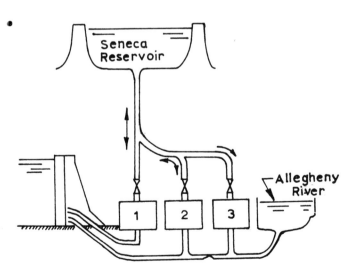

Figure 50. Plan and elevation of Seneca pumped storage plant near Warren, Pennsylvania.

overall efficiencies of the Seneca plant is reported[2] to range between 71% and 79%.

The storage area required for a given generating capacity is dependent on the height of the water storage. For 100-foot pumped height

$$G = \rho h = 62.4 \ (lb/ft^3) \times 100 \ ft = 6240 \ ft\text{-}lb/ft^3 = 0.00235 \ kwh/ft^3$$

Based on this value, Table 11 gives size and cost figures for pumped energy storage in the range of 200 to 60000 kwh. Using the 75% overall efficiency figure the unit cost for a 60000 kwh storage facility (last line Table 11) is:

$$\text{Unit Cost} = 0.0131/0.75 = \$0.0175/kwh$$

Table 11. Pumped Water Storage Costs.

Storage Capacity kwh	Volume cu ft	Unit Cost ¢/ft³	Total Cost $	40-year cost $/kwh
200	85,000	75	63,800	0.0218
2,000	850,000	60	510,000	0.0175
20,000	8.5×10^6	50	4,250,000	0.0146
60,000	25.5×10^6	45	11,500,000	0.0131

This figure does not include the cost of the reversible pump/turbine and motor/generator unit, which for a 60000-kwh storage facility would add $0.0002/kwh (amortized over 40 years) to give

$$\text{Unit Cost} = \$0.0177/kwh$$

The major environmental concern for pumped water storage systems is the drawdown, which takes place during power generation. This results in an unsightly exposure of the reservoir and prohibits use for recreational purposes.

Compressed Gas Storage

For gas compression systems, air is the most attractive working fluid. Helium is better theoretically, but a low-pressure tank for recovery is necessary for helium (or any other gas than air), and the cost of the tank far outweighs any theoretical advantage in the compression process.

An air storage peaking power plant is illustrated in Figure 51. A conventional gas turbine modified so the turbine and compressor may be uncoupled and operated separately, and a flow control valve to a large air storage underground are the principal features of the system. During off-peak low-load periods, the turbine coupling is dis-

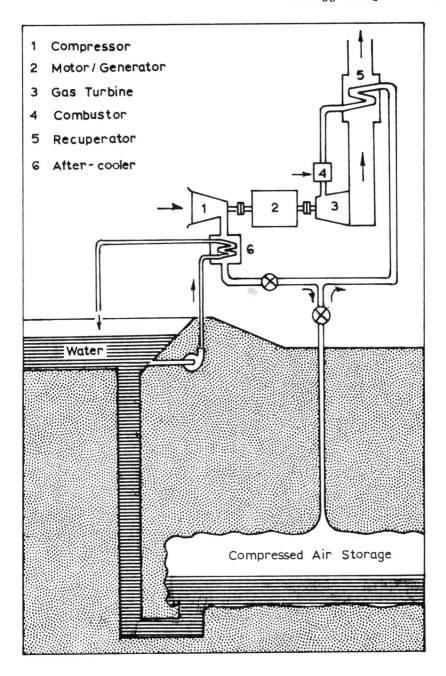

1 Compressor
2 Motor / Generator
3 Gas Turbine
4 Combustor
5 Recuperator
6 After-cooler

Water

Compressed Air Storage

Figure 51. Air storage gas turbine power plant.[3]

engaged and the compressor is driven by the electric generator operating as a motor, and taking power from the grid. During peak periods, the compressor clutch is disengaged, and the compressed storage air is returned through the recuperator to the burner and turbine, which now drives the generator for power delivery to the grid.

The least expensive underground storage areas include dissolved out salt caverns, porous ground reservoirs, depleted gas and oil fields, and abandoned mines. Such areas are usually hard to find near major electric load centers. Caverns formed by nuclear explosives may be made available in such cases. The requirement is for a geological formation that will fracture to a sizable volume and porosity by the nuclear explosion. Rock formations of granite, shale, or limestone are particularly suitable.

A type of air storage system formed by nuclear explosion is shown in Figure 52. To avoid large pressure fluctuations either hydrostatic pressurization or very large air storage volumes are needed. The hydrostatic pressure system, as illustrated in Figure 51 with its attendant small storage volume would appear most attractive from a cost standpoint. There are complications, however, which increase the expense. First, a water storage reservoir is required and must be of about the same volume as the air cavity. Second a large conduit extending to the bottom of the cavity is required. (A 20-foot diameter conduit was used in a Swedish study.[3]) The large diameter minimizes pressure drop losses associated with transporting the water in and out of the cavity. Third, the distance between the reservoir

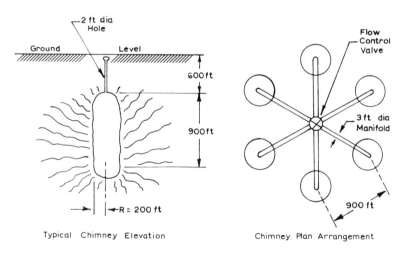

Figure 52. Air storage formed by nuclear explosion.

water surface and cavity water surface must remain nearly constant to minimize fluctuations. Fourth, the cavity air pressure resulting from the hydrostatic head must match the pressure required by the gas turbine components.

Cavities produced with nuclear explosives usually approximate rubble-filled hemisphere-capped cylindrical chimneys with the height about 4.5 times the chimney radius as indicated in Figure 52. This type of geometry would result in an expected reduction of 60% in cavity pressure over the duty period. Further, a 100-kiloton nuclear explosive results in a minimum head of about 660 lb/in^2 or a pressure ratio of 45 to 1.[4] This is not compatible with industrial gas turbine components in the USA which typically operate on a 10 to 1 pressure ratio.

On the other hand, nuclear explosives appear to be particularly appropriate for air storage systems of large volumes. Gas turbine components should be able to tolerate pressure fluctuations of ± 5% of the design pressure without serious decrease in performance. This fluctuation maximum can be obtained with an air storage volume 10 times that of a hydrostatically pressurized system. Large volume storage has the additional advantages of eliminating problems and expense of the hydrostatic pressurization system and allows for operation beyond design duty period by increasing the allowable pressurization system.

A detailed study of costs for a 220 Mw air storage gas turbine power plant has been performed by E. K. A. Olsson of the Stal-Laval Turbine Co. in Sweden.[5] Technical data and specific cost breakdown are given in Table 12. The efficiency is 71% or a heat rate 4770 B/kwh. When used in the conventional manner the compressor requires 161 Mw leaving a net output of 73 Mw with an efficiency of 27% or heat rate of 12,600 B/kwh.

In the area studied the total above ground costs are estimated at $39/kw for a turnkey job. Underground the costs are estimated at $11/kw for a total of $50/kw. This specific cost of the plant is 70% lower than a conventional gas turbine plant to give an equivalent peaking capacity output, and is said to be generally lower than that for pumped water storage.

Batelle Pacific Northwest Laboratories has also conducted a detailed study of air storage peaking power plants using large-volume cavities created by nuclear explosives. Table 13 summarizes the findings of this study.[4] The calculations for the air storage plant are based on low air storage temperature (120°F), and rock volumes which yield high void volume per kiloton of nuclear explosive. On this basis it is clear from Table 13 that air storage systems are clearly competitive with other peaking systems.

Table 12. 220-Mw Air Storage Power Plant.[5]

A. Technical Data

Air flow to storage	775 lb/sec
Maximum air storage pressure	640 lb/in^2
Air storage temperature	59 F
Cavern depth	1425 ft
Cavern volume	970,000 ft^3
Compressor power	161 Mw
Continuous power	73 Mw
Efficiency/heat rate	
1) Peak-load operation	71% at 4770 B/kwh
2) Continuous operation	27% at 12650 B/kwh

B. Cost data

Mechanical and electrical equipment, building, foundation, and fuel	
building, foundation, and fuel storage	$31.8/kw
HV transformer	4.3
Administration (2%) and interest (7%)	2.9
Subtotal above ground	$39.0
Cavern for 5-hr operation, and hydrostatic pressurization system	8.6
Adminstration and interest	2.4
Subtotal for below ground	11.0
Total excluding land, roads and taxes	$50.0/kw

Table 13. Comparison of Costs for Various Peaking Systems.[4]

Type-System	Size (Mw)	Capital Cost $/kw	Total Cost mills/kwh
Pumped hydro	7.2–600	75–125	9.71–13.82
Gas turbines	58(1973)[a]	77.6	11.96–12.68
	250(1980)[b]	66.5	10.05–10.70
Steam peaking	135–400	105	14.57
Air storage	175(1973)[a]	65.7	9.38
	542(1980)[b]	42.2	7.60

[a]Based on 1973 gas turbine technology.
[b]Based on 1980 gas turbine technology.

FLYWHEELS

Flywheels have been used for decades for regulating the speed and uniformity of motion of machines. As energy storage devices, flywheels are particularly attractive because of their high energy and power densities. High-strength composite flywheels are being studied for applications ranging from vehicles to power plants. In both applications, the flywheel could be charged during time periods when power demands are low and the stored energy made available when demand is high. In the operation of a power plant, flywheels could serve the same purpose as pumped water or compressed gas energy storage discussed previously.

For vehicular applications, it has been estimated[6] that an energy-storing flywheel, coupled to an internal combustion engine, could reduce the size of the engine which the hybrid system replaces by 50%. Another particularly attractive feature of a flywheel system is regenerative braking whereby energy, normally dissipated as heat in the brakes, could be returned to the flywheel. In certain applications such as local shuttle transportation, high-performance flywheels alone could power low-demand vehicles, being recharged at points along the route.

Energy Storage

The ratio of kinetic energy to weight, that is, the specific energy, which is stored in a rotating flywheel is given by the expression

$$E = \frac{KE}{wt} = \frac{Mv^2/2}{wt} \tag{50}$$

where M is the mass of the flywheel and v is the tangential velocity measured at some distance, r, from the center of the wheel. Since the weight of the flywheel is Mg (g = acceleration of gravity, 32.17 ft/sec^2) and the flywheel has some rotational velocity, ω (radians/unit time), Equation (50) can be rewritten as

$$E = \frac{Mr^2\omega^2/2}{Mg} = \frac{r^2\omega^2}{2g} \tag{51}$$

where the tangential velocity at a point on the flywheel is related to the rotational velocity by the relationship $v = r\omega$. It is easy to imagine that as the rotational velocity of the flywheel increases, stresses in the flywheel, both radial and tangential (hoop), will increase. For the simple case of a thin-rim flywheel, it can be shown that the tangential stress increases as the square of the velocity. Such an analysis gives the following relationship:

$$\omega^2 r^2 = \frac{g\sigma}{\rho} \quad . \tag{52}$$

If σ is taken as the maximum design stress and ρ is the density of the flywheel material, ω given in Equation (52) is the maximum rotational velocity for a wheel or radius r below which the design stress will not be exceeded. Combining Equations (51) and (52) yields a relationship between the specific energy density for a thin-rim flywheel and the material properties, strength, σ, and density, ρ:

$$E = \frac{1}{2} \frac{\sigma}{\rho} \qquad (53)$$

The proportionality constant, 1/2, is characteristic of a thin-rim geometry, whereas a generalized expression is

$$E = K_s \frac{\sigma}{\rho} \qquad (54)$$

where E is specific energy in $\frac{\text{in-lb}}{\text{lb}}$

K_s is the flywheel shape factor (dimensionless)

σ is the material design stress level in lb/in^2

ρ is the material density in lb/in^3

The shape factor for several flywheel geometries are;[6] flat unpierced disc ($K_s = 0.606$), single filament bar ($K_s = 0.333$), flat pierced disc ($K_s = 0.305$), compared to $K_s = 0.500$ derived previously for a thin-rim flywheel.

In addition to specific energy, consideration of the volumetric efficiency of a flywheel design is necessary. This is particularly true for vehicular applications. The volumetric efficiency refers to the relationship between the static volume of the flywheel and the swept volume. For example, the volumetric efficiency of a flat pierced disc is considerably higher than for a single filament bar even though they have comparable shape factors. The so-called superflywheel, a rimless, multispoke flywheel of which much has been written in recent years, has a relatively low volumetric efficiency. On the other hand, concentric-ring designs, where the rings of fiber composites are separated by small gaps filled with resilient bonding materials, make efficient use of the wheel's volume.

Flywheel Materials

Numerous materials for flywheel applications have been studied. Several of them, along with their important properties, are listed in Table 14. The recommended working stress represents an assumed design life of the flywheel of 10 million cycles. Normalized costs are based on mill-run quantities and were calculated as follows: material cost was divided by the working stress-to-density ratio and then divided by the resulting value for the least expensive material. Thus, the normalized cost represents the cost for each material, (relative to the least expensive), to provide an equivalent energy storage capability for a given flywheel configuration.

The first material shown, 18Ni–400 maraging steel, is one of the highest strength steels currently available in mill-run quantities. Its relatively high cost is due largely to its high nickel content.

AISI 4340 steel is a relatively common high-strength hot-rolled steel that has an attractive combination of characteristics. This steel contains only small amounts of alloying elements (1.8 w/o Ni and 0.8 w/o Cr) and its relatively low cost in combination with a high recommended working stress makes it an excellent candidate material for low-cost flywheels.

The plain carbon steels, 1040 and 1020, and cast iron are the least expensive materials shown, but their low working stress levels result in less attractive normalized costs.

Table 14. Flywheel Materials.[a]

Material	Density, ρ (lb/in³)	Ultimate Tensile Strength (ksi)	Yield Strength (ksi)	Rec. Working Stress, σ (ksi)	$\frac{\sigma}{\rho}$ (x 10⁶)	Matl. Cost ($/lb)	Norm Cost
18Ni–400 (Maraging Steel)	0.289	409	400	260	0.900	2.25	5.3
4340 Steel	0.283	260	217	130	0.459	0.60	2.78
1040 Steel	0.283	87	58	36	0.127	0.30	5.00
1020 Steel	0.283	68	43	25	0.088	0.30	7.23
Cast iron	0.280	55	37	20	0.071	0.30	8.94
2024-T851 (aluminum alloy)	0.100	66	58	35	0.350	0.50	3.03
"E" glass	0.075	200	—	67	0.890	0.42	1.00
S-1014 glass	0.072	260	—	87	1.210	0.74	1.31

[a]After Lawson.[6]

The high-strength wrought aluminum alloy, 2024, nominally contains 4.5 w/o Cu, 0.6 w/o Mn, and 1.5 w/o Mg. This is a heat-treatable alloy; that is, its properties can be greatly improved by selective heat treatments. T 851 temper indicates a processing treatment consisting of solution heat treatment, cold working, artificially aging (strengthening) and stress relieved by stretching.

The two nonmetallic materials, E-glass and S-glass have similar compositions ($-$54 w/o SiO_2 + 14 w/o Al_2O_3 + 16 w/o CaO + 10 w/o B_2O_3 + w/o MgO), but different properties by virtue of different processing techniques. Although their normalized costs are low, their low densities result in poor volumetric efficiencies in flywheel designs based on current technology. Other significant development problems for glass-epoxy flywheels include attachment and balancing.

The choice of a flywheel geometry-material combination for a given application will be based on requirements particular to that application. Of course, in many instances, cost may be the determining factor, while in others swept volume or specific energy may be the determining factors. Analyses of flywheel-internal combustion engine hybrids have been made and the results compared to electric battery-i.c. engine combinations. Lead-acid batteries, to be discussed in the next section, have specific energies up to 10–15 watt-hrs/lb and several flywheel geometry-material combinations have comparable energy densities if they are operated at a sufficiently high material working stress-to-density ratio, σ/ρ. Typical values of specific energy for several flywheel geometries are given in Table 15.

Table 15. Specific Energy ($\frac{\text{watt-hr}}{\text{lb}}$)[a]

Geometry	Material Working Stress-to-Density Ratio	
	$10^5 \frac{\text{in-lb}}{\text{lb}}$	$10^6 \frac{\text{in-lb}}{\text{lb}}$
Flat unpierced disc	4.0	20
Thin-rim	1.7	17
Single filament bar	1.1	10.5

[a]After Lawson.[6]

BATTERIES

Batteries have become an extremely useful source of energy. This is because it is generally difficult to store energy in a form for convenient later use. The electric storage battery, however, can provide relatively long storage of electrical energy in convenient, reliable, and portable form.

The work of Galvani in 1790 is generally credited with being the pioneer work on electrochemical cells. Volta in 1810 and Daniell in 1836 are other early workers in this field who have left their names associated with their work. However, batteries suitable for ordinary applications came somewhat later. The "dry" cell and the "lead-acid" battery evolved around 1859–1866.

Primary Batteries

Batteries may be classified according to whether or not they are rechargeable. Those not rechargeable are termed primary batteries. These batteries are manufactured for one-time use and subsequent disposal. The familiar flashlight battery is predominant among primary cells. It is more accurately called a Leclanché dry cell. It consists of a carbon electrode, a zinc can and an electrolyte which is predominantly manganese dioxide but which contains small amounts of additional compounds. In 1963, improvements made possible from 4–18 times the capacity at some increase in cost. Later, a still more expensive cell was developed, known as the alkaline-manganese cell. This cell has generally the best performance of all Leclanché type cells. Dry cells are suitable energy sources only where electrical energy is needed in small amounts and where the flexibility and portability characteristics are paramount.

Other primary cells have been developed for special purposes. The magnesium-seawater cell is a battery which is stored dry and which uses seawater as the electrolyte. This battery can endure long storage in the dry form and when energized by addition of seawater can provide high output and a high rate of discharge, making it particularly suitable to marine applications. Equipment such as torpedos and emergency radio generators often are provided with this type battery. Other very high performance and costly batteries are under development for military applications. Foremost among these types are silver or zinc and air batteries.

Lead-Acid Storage Battery

The secondary or rechargeable battery is the energy source of greatest significance to all but the very specialized applications. Predominant among secondary batteries is the familiar lead-acid battery. Its use exceeds its nearest rival by at least a factor of 10. The major application is as a starting battery for automotive engines. The advantages of the lead-acid battery are a) capability of supplying high or low currents over a wide range of temperatures, b) capability of relatively long-term storage, particularly in the dry-charged condition, c) capability for undergoing many (hundreds) cycles of charge and discharge, and d) relative cheapness when compared to potentially competitive batteries.

A chemical reaction occurs in an electrochemical battery which can provide for the flow of electrical current through an external circuit. This electrical current consists of a flow of electrons, each carrying an electrical charge. The amount of energy available with each charge is the electrical potential or voltage. The work which can be done by the system is

$$dW = Vdq \qquad (55)$$

This work is reversible (and maximum) when the voltage V is maximum. This occurs when the flow of charge q is zero and the voltage then is known as the electromotive force or emf. As the current increases the value of the voltage at the terminals decreases due to the internal resistance of the battery and correspondingly less work can be done. The work done by a battery operating reversibly is

W (joules) = V (volts or joule/coulomb) x q (coulombs)
$\dot{W}$ (watt or joule/sec) = V (volts) x I (amperes or coulombs/second)

The chemical reaction can best be studied by considering the production of electrons or unit charges. The charge carried by N_A monovalent ions where N_A is Avogadro's number is one Faraday. Avogadro's number N_A = 6.025 x 10^{23} atoms/gm atom or molecules/gm mole). A more general way of stating this is to say that the charge carried by one gram-equivalent weight of ion is a Faraday. A gram equivalent weight is the atomic or molecular weight divided by the valence; that is, for example, ½ gram atomic weight for a divalent ion.

1 Faraday = 26.81 amp-hr = 6.025 x 10^{23} unit charges.

The total charge is the number of Faradays times the charge of a Faraday or

$$q = n\mathcal{F} \tag{56}$$

In terms of work for a reversible reaction:

$$W = Vq = n\mathcal{F}V \tag{57}$$

Because of its importance in energy applications we will consider the lead-acid battery reaction. The electrolyte in a lead-acid battery is sulfuric acid, in which the following reaction takes place:

$$H_2SO_4 \rightleftarrows 2H^+ + SO_4^{--}$$

The positive plate is lead oxide in the charged state. It reacts with the electrolyte as follows:

$$PbO_2 + SO_4^{--} + 4H^+ + 2e \underset{charge}{\overset{discharge}{\rightleftarrows}} PbSO_4 + 2H_2O$$

and produces a potential of +1.685 volts. The electrons are produced during discharge as the lead goes from a +4 valence to a +2 valence. The negative plate is lead in the charged state. It reacts with the electrolyte as follows:

$$Pb + SO_4^{--} \underset{charge}{\overset{discharge}{\rightleftarrows}} PbSO_4 + 2e$$

and produces a potential of −.356 volts. The overall reaction, obtained by adding the above equations for the simultaneous reactions is

$$PbO_2 + Pb + 2H_2SO_4 \underset{charge}{\overset{discharge}{\rightleftarrows}} 2PbSO_4 + 2H_2O$$

from which it can be seen that as the reaction proceeds in the discharge direction, the electrolyte is converted in part from sulfuric acid to water. The difference in specific gravity between these liquids allows a determination of the degree of charge of the battery through a simple specific gravity measurement.

Observation of the reaction equations indicate that for 1 Faraday of current flow ½ gram mole of positive plate, ½ gram mole of negative plate and one gram mole of electrolyte is needed. Therefore, 297/2 grams of lead oxide, 207/2 grams of lead, and 98 grams of electrolyte are needed. The theoretical weight of battery needed for a given capacity can therefore be calculated. Actual batteries weigh considerably more because of need for structure and due to problems associated with unreacting elements.

Nickel-Cadmium Batteries

Nickel-cadmium batteries have many advantages when compared to lead-acid batteries. They have excellent life, being capable of thousands of charge-discharge cycles and, in addition, they can be stored for long periods without damage in either the charged or discharged state. These batteries can provide high current density on discharge with little decrease in voltage and can be recharged rapidly without damage. They are expensive, however, so their use is mostly limited to small size applications, where rechargeability is desirable and where the cost increment is acceptable. No possibility exists for widespread use of high-capacity nickel-cadmium batteries because of the limited amount of the raw materials available for their manufacture.

Charging and Discharging

All batteries have similar curves of charge-discharge as illustrated in Figure 53. The voltage difference on charge versus discharge is due to internal losses. The area under the curves at the same value of current are proportional to the energy. As current increases, the difference between voltage and emf increases, both during charge and discharge. At useful charge and discharge rates, considerably more energy must be added to the battery during charge than is available during discharge. The nickel-cadmium battery has a more favorable voltage curve than lead-acid batteries.

The characteristics of the types of battery cited vary widely. Even within a particular type of battery considerable variation, based on quality of manufacturer, is possible. Applications for batteries are equally varied and the selection of a battery for a particular application must be made first to meet technical needs, but second to meet cost considerations. For example, the nickel-cadmium battery costs about 10 times as much as a lead-acid battery and 20 times as much as a dry cell for the same electrical capacity. It is attractive for many applications, but certainly not for all, simply because of its cost. Batteries meet an important need which could, in total, not be well met by any other means.

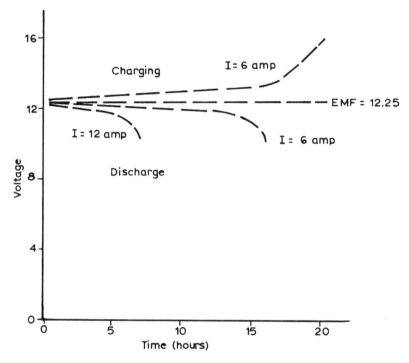

Figure 53. Typical charge and discharge curves for a 6-cell lead-acid battery of 100 amp-hr at 6-amp rate capacity.

The Sodium-Sulfur Storage Battery

The development of a battery based on sodium and sulfur was begun by the Ford Motor Company in 1963. This battery system, as we will see, is of particular interest because of the high power density or specific power (watts/pound) as well as the high energy density (watt-hours/pound) which such batteries can achieve. It is currently believed that sodium-sulfur batteries will be capable of power densities of up to about 100 watts/pound. For comparison, Figure 54 shows the specific power (watts/pound) versus specific energy (watt-hours/pound) for a variety of batteries and engines. The tremendous superiority of sodium sulfur as compared to lead-acid batteries is obvious. It is also evident, however, that even sodium sulfur batteries are somewhat inferior to internal combustion engines.

Sodium-sulfur batteries would still be able however, to give driving ranges of up to nearly 500 miles under optimum conditions or of up to nearly 300 miles routinely. Such driving ranges would go a very long way toward making electric automobiles truly practical.

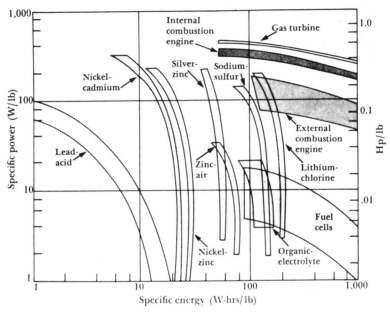

Figure 54. A comparison of the specific power (watts/pound) versus specific energy (watt-hours/pound) for both batteries and combustion engine systems.[7]

Figure 55 shows the schematic arrangement of a sodium-sulfur battery. It will be noted that whereas a lead-acid battery consists of solid electrode plates and a liquid electrolyte, the sodium-sulfur battery consists of liquid electrodes (one of liquid sodium and one of liquid sulfur) and a *solid* electrolyte. During the discharge of this cell, the following overall reaction occurs:

$$2Na + S \rightarrow Na_2S$$

The sodium is the negative terminal of this cell and at this electrode the sodium is oxidized according to the reaction

$$2Na \rightarrow 2Na^+ + 2e^-$$

The sulfur is the positive terminal of this battery and at this electrode the sulfur is reduced according to the reaction

$$2Na^+ + S + 2e^- \rightarrow Na_2S$$

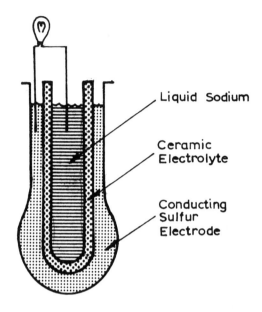

Figure 55. A schematic of a sodium-sulfur battery.

The melting point of Na_2S is 920°C, but if sulfur is present in excess, this Na_2S further reacts according to

$$Na_2S + 4S \rightarrow Na_2S_5$$

The melting point of Na_2S_5 is 251°C and the sulfur electrode therefore remains liquid as long as the temperature is above this value. For this battery to operate, it is necessary to have a ceramic electrode material which will not pass electrons but which will pass sodium ions (Na^+). It was the discovery by materials scientists, that the substance beta alumina has these properties, which has stimulated virtually all present interest in the sodium-sulfur battery. In beta alumina, which has the nominal composition $Na_2O \cdot 11\ Al_2O_3$, the sodium ions form a loose bridge structure between so-called Al_2O_3 "blocks." These sodium ions are only very loosely bound to the Al_2O_3 blocks and have a very high mobility, *i.e.*, the sodium ions are free to move easily in this structure. There are no free electrons in beta-alumina and so it would normally be thought of as an electrical insulator. However, charge can be transported by the Na^+ ions. A membrane made of beta alumina therefore allows the movement of Na^+ ions from the sodium to the sulfur but prevents the movement of electrons.

Beta alumina is only one member of a general class of materials which have the formula $A_2O \cdot n \, M_2O_3$ where A can be Na, Li, K, Rb, Ag, or Tl, M can be Al, Ga, or Fe, and n can have values between 5 and 11. Such materials could be used to produce a great variety of new batteries, such as lithium-sulfur, lithium-selenium, or lithium-thallium batteries. It can be readily seen that the existence of this new class of ceramic electrolyte materials opens a very wide field for battery research. Of course, there are many problems which must be dealt with, as for example, the high temperature needed for the sodium-sulfur cell to operate, and also the fact that metallic sodium can, in certain cases, diffuse along beta alumina crystal boundaries and cause the battery to short circuit. Nevertheless, the sodium sulfur battery is currently one of the most promising areas for the successful development of practical electric automobiles.

REFERENCES

1. "Electric Power in the Southeast 1970-1980-1990," A report to the Federal Power Commission by The Southeast Regional Advisory Committee, April 1969, pp. 2–7; 2–16; 2–17.

2. Fitzgerald, J. F., E. A. Cooper and F. P. Solomon. "Operation of Seneca Pumped Storage Plant," *IEEE Trans.* **PAS–92**, No. 5, 1510 (1973).

3. Harboe, H. "Economic Aspects of Air Storage Power," *ASEA J.* **44**, No. 2, 43–47 (1973).

4. Fryer, B. C., "Air Storage Peaking Power Plants—Utilizing Modified Industrial Gas Turbines and Cavities Created with Nuclear Explosives," Report BNWL-1748, May 1973, Battelle Pacific Northwest Laboratories, Richland, Washington.

5. Olsson, E. K. A. "220 MW Air Storage Plant," ASME Paper 70-GT-34, 1970, Amer. Soc. Mech. Eng., United Engineering Center, New York, N.Y.

6. Lawson, L. J. "Design and Testing of High Energy Density Flywheels for Application to Flywheel/Heat Engine Hybrid Vehicle Drives," *1971 Intersociety Energy Conversion Engineering Conference*, Boston, Massachusetts, August, 1971, pp. 1142–1150.

7. National Air Pollution Control Administration. "Control Techniques for CO, NO_2, and Hydrocarbon Emissions from Mobile Sources" (Washington, D.C.: U.S. Government Printing Office, 1970).

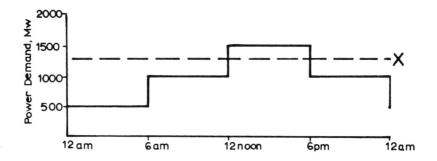

PROBLEMS

1. The above sketch shows an idealized power demand cycle for a community. The peak load is 1500 Mw for 6 hours (12 noon–6 p.m.). The minimum load is 500 Mw for 6 hours (midnight–6 a.m.). The other 12 hours have a 1000–Mw power demand. It is proposed to have a pumped water storage facility for meeting peak demand. The overall efficiency of the storage facility (*i.e.*, electrical energy delivered during high demand to electrical energy drawn during low demand) is 72%. This breaks down to 85% pump/motor and 85% turbine/generator efficiencies.

   In the above diagram X represents the capacity of the steam power plant facilities. The power requirement to be met above X is the storage plant capacity: a) Determine the required installed capacity of the steam power plant (in Mw); b) Determine the peak power output of the storage plant (Mw); c) Calculate the volume of water reservoir required for a 100-foot height above the turbines neglecting any piping or valve losses; d) What are the major advantages and disadvantages of such a pumped water storage facility?

2. The engine for a vehicle which develops an average of 100 hp during 5 hours of operation between fill-ups is being replaced by a flywheel. Determine the weight of the flywheel required if it is to be a flat unpierced disc constructed of 4340 steel.

3. Flat unpierced, cast iron flywheels are being considered for storing energy at a power plant to meet peak loading needs. Assume that the hub diameter is 10% of the total wheel diameter and the thickness of the hub is twice the thickness of the rim. a) Determine the flywheel diameter and weight required to store 100 kwh of energy; b) Show whether the total weight of cast iron required for 10 flywheels having the same total energy capacity would be less than, more than, or the same as that required for the single flywheel in part (a).

4. What principal characteristic differentiates a battery from a fuel cell?
5. If a voltage equal to its emf is applied across the terminals of a lead-acid battery, will the battery become fully charged in due course?
6. How much work is done when a lead-acid battery is discharged at a constant 6.0 volts, passing 2 Faradays of charge? How much lead-oxide enters the reaction? How much work could be done if the process were reversible?
7. How would you discharge a lead-acid battery to get the maximum amount of work from the process?

Chapter 7

Energy for Transportation

INTRODUCTION

At the present time, fueling the transportation industry in the United States accounts directly for about 24% of the total energy used annually. This percentage represents direct fuel costs only. It has been estimated that, if manufacturing and service costs are included, this figure is closer to 50%. Furthermore, the percentage of transportation energy derived from petroleum is about 95.5%, and most projections show that this percentage will either remain steady or increase slightly by the year 2000. In view of our limited petroleum resources, discussed in Chapter 2, the overall supply-and-demand of energy for transportation should be carefully examined.

In this chapter, we begin with an examination of transportation energy needs. The energy requirements for both passenger and cargo transportation systems will be considered. Then, we will look closely at the automobile and consider some alternatives to the present situation.

TRANSPORTATION ENERGY USAGE

Table 16 shows the almost total dependence of the transportation industry on petroleum energy supplies. This dependence is accounted for by the fact that our primary power propulsion systems are the internal combustion engine (gasoline and diesel) and the gas turbine. The table shows that by the year 2000, about 70% of the petroleum consumed in the United States will be used for transportation. If such a trend were to continue, by the year 2020 the U.S. transportation oil demand would equal the world's petroleum production. The relationship between U.S. transportation energy demands and world oil production is summarized in Figure 56.

Since most of the energy used for transportation is derived from petroleum, it is instructive to look at the distribution of energy within the transportation sector. This information is shown in Table 17.

139

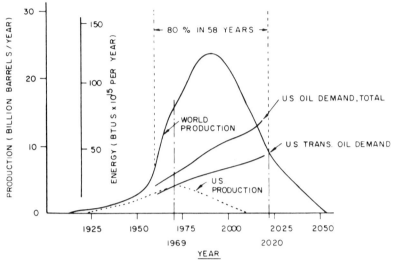

Figure 56. United States energy demands on oil production.[2]

Table 16. Petroleum Consumption in the United States[a]

Year	Total Consumption 10^{15} Btu	Percentage Used For Transportation	Percentage of Transportation Energy From Petroleum
1950	13.49	50.3	77.8
1955	17.52	52.0	92.0
1960	20.07	51.7	95.3
1965	23.24	52.5	95.5
1970	29.62	53.2	95.5
1980	35.98	57.6	96.1
2000	57.60	72.3	97.1

[a]From Hirst.[1]

These data show that automobiles and aircraft accounted for 33.1% of energy requirements for intercity passenger travel. Yet, in terms of propulsion efficiency (net passenger mpg), airplanes and autos are the two least efficient modes of passenger transport. Airplanes and autos respectively require eight and four times as much energy per passenger mile as buses. In spite of this, trains, which have an even higher propulsion efficiency than buses, and buses accounted for only 0.6% of the energy consumed in intercity travel in 1970.

The present situation for urban transportation is even more energy-intensive. Not only does the personal automobile account for nearly

Table 17. Distribution of Energy Within the Transportation Sector[a]

	Percent of Total Energy	
	1960	1970
Automobiles		
Urban	25.2	28.9
Intercity	27.6	26.4
Aircraft		
Freight	0.3	0.8
Passenger	3.0	6.7
Railroads		
Freight	3.7	3.2
Passenger	0.3	0.1
Trucks		
Intercity freight	6.1	5.8
Other uses	13.8	15.3
Waterways		
Freight	1.1	1.0
Pipelines	0.9	1.2
Buses	0.7	0.5
Other	17.3	10.1
Totals	100.0	100.0
Total Transportation Energy		
Consumption (10^{15} Btu)	10.9	16.5

[a]From Hirst.[1]

all urban passenger travel, the efficiency of fuel usage is substantially lower than for intercity travel. Studies have shown that as much as 47% of the total driving time is spent either decelerating or idling, with the balance accounting for acceleration and cruising. Of course, during periods of idling and deceleration, the efficiency (work out/energy in) is zero.

Some Solutions or Nonsolutions

The foregoing brief analysis of transportation energy usage coupled with information concerning energy resources in Chapter 2 provide ample evidence that current transportation trends cannot continue. Many solutions to the transportation energy problem have been proposed. Among the more common proposals are use of mass transit systems, substitution of bicycling and walking for short auto trips (<2.5 miles), and utilization of primary energy sources other than petroleum. Some of the solutions proposed may be useful in the intermediate time range, some in the long term, and others may not work at all.

The attractiveness of mass transit, and bicycling and walking, stems from the fact that their energy efficiencies are much higher than for autos and airplanes. This is illustrated for urban travel. These data show that buses are about four times as efficient as autos while bicycles are better than twenty times as efficient. Utilization of such transportation modes would certainly improve our overall national propulsion efficiency. However, one must be cautious when considering such data! The energy requirement shown for buses is based on an average occupancy of 20 passengers. Such occupancy rates can probably be realized only on high population density routes and during "prime" time. Thus, while most of our larger cities presently have bus systems, extending these systems into the suburbs may not contribute measurably to a reduction in total transportation energy needs. Likewise, the almost 30-fold energy savings suggested by Table 18 for bicycling is probably misleading. Considering that bicyclers will require more food, and the amounts of primary energy involved in producing, transporting, and marketing that food, the real energy savings of bicycling over automobile riding may be only a factor of two to three.

Table 18. Energy Efficiency for Urban Passenger Traffic.[a]

Mode	Energy Requirement[b] Btu/(passenger mile)
Bicycles	180
Walking	300
Buses	1240
Automobiles	5060

[a]From Hirst.[1]
[b]Efficiencies for walking and bicycling assume 893 Btu/hr required for moderate output, 5 mph for bicycling and 3 mph for walking.

THE AUTOMOBILE

The automobile is at the heart of our increasing transportation energy problem because of the extent to which our society and economy are dependent on it. The following statistics, taken from a recently published study[3] sponsored by the RANN Program of the National Science Foundation, serve to illustrate the extent of this involvement:

"In January 1970 there were 87 million passenger cars registered in the United States.

One-hundred-eight million drivers drove these cars 859 billion miles over 3.7 million miles of roads.

> Sixty-three billion gallons of fuel were consumed in 1969 to provide the energy for this mobility.
>
> Domestic new cars were manufactured by four companies in 1850 establishments spread across 17 states and 40 cities.
>
> New car manufacturing employed 700,000 workers.
>
> The automobile industry is twice the size of the next largest manufacturing industry and constitutes ten percent of our Gross National Product (GNP).
>
> Adding in related industries (transportation, sales and service, petroleum refinery, etc.) the automobile is responsible for 13 million jobs — about one-sixth of total U.S. employment — and about one-fourth of all retail sales."

Considering that Americans, mobile, independent, and increasingly affluent, will continue to move from the inner city to the suburbs, it is probably unrealistic to expect a decrease in the importance of the automobile in our society.

The figures cited above showed dramatic increases over the two-decade period from 1950–1970. Intercity passenger traffic increased from 510 billion passenger miles in 1950 to 1180 billion in 1970 and the automobile provided 87% of those miles. During the same period urban passenger traffic increased from 388 to 987 billion passenger miles and the percentage supplied by the automobile rose from 89.6 in 1950 to 95.4 in 1970. (Statistics from Hirst.[1]) The latter increase reflects a decline in utilization of buses, a mode of travel which we have already shown to be much more efficient than the automobile.

At the same time that more autos are carrying more people more miles, until recently they have been doing it less efficiently. There are several reasons for this. The most publicized factor is the requirement for exhaust emission control equipment. Estimates of the penalty resulting from 1976 standards range from a high of 30% increase in fuel consumption to a concensus estimate of 18–20%. The requirement for a 10-mph bumper to meet increasingly stringent damageability and safety regulations is another example of a factor contributing to increased fuel consumption. Although insurance and repair bills may be somewhat reduced, strengthening the frame and adding such bumpers generally will add well over 100 lb to the vehicle weight and a concomitant increase in power and fuel requirements.

The gasoline shortage in early 1974 dramatically illustrated the problems we are facing with respect to transportation energy. To the consternation of the environmentalists, the federal imposition of pollution and safety standards, which are so counterproductive to efficient utilization of energy, are being seriously questioned. On the other hand, the consuming public has initiated some beneficial pattern changes, probably as a result of the gasoline shortage scare.

Probably the most significant change resulting from the 1974 gasoline shortage is the switch to lighter weight automobiles. Automobile weight is the single most important determinant for fuel economy. Figure 57 shows the relationship between automobile weight and energy intensiveness. The data do not reflect other determinants of fuel economy such as automatic transmissions and horsepower-to-weight ratios. Nevertheless, a large scale switch from heavy to lighter vehicles could significantly improve the fuel consumption picture. It is probably too soon to tell whether improvement in fuel consumption (probably only a reduction in the rate of increase) due to a switch to lighter vehicles and other factors, such as nationally imposed speed limits, will continue or just how significant they will be.

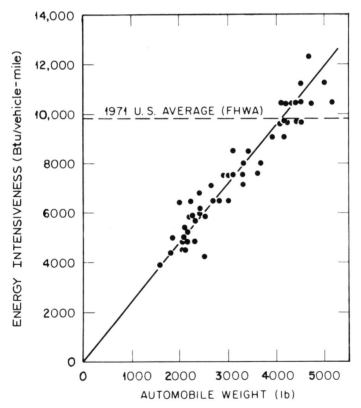

Figure 57. Automobile fuel consumption as a function of vehicle weight for 1971- and 1972-model cars.[4]

Other marginal impacts on the transportation energy picture may result from development studies of both internal and external combustion engines operating on cycles other than the standard Otto cycle (today's IC engine). Such studies are being conducted by many companies and laboratories. Typical thermal efficiencies of several of these propulsion systems is shown in Table 19.

Table 19. Thermal Efficiency of Power Systems.

Propulsion System	ηth(%)
Gasoline/piston (Otto)	22–32
Gasoline/rotary	18–30
Diesel	35–38
Gas turbine (regenerative)	20–30
Stirling	29–40

The peak efficiency of some of these engines is significantly higher than that of the conventional gasoline/piston engine. However, trade-offs would be required if a substitution were to be made. For example, both the diesel and Stirling engines are heavy and bulky compared to a standard IC engine of comparable performance and in addition the diesel engine is considerably noisier. With all factors considered, these alternatives offer, at best, to only prolong the problem of decreasing fossil fuel supplies.

No matter what is done, eventually the desirable petroleum fuel resources will be depleted! Even hydrocarbon fuels derived from coal and oil shale are not unlimited. In view of the fact that the technologies for coal gasification and extraction of oil from oil-bearing shales have not been fully developed and that the comparative economics of such ventures cannot yet be accurately projected, other solutions to the future transportation energy problem must be considered.

Two alternatives which are receiving considerable study nationally will be looked at in some detail. First, a renewable energy resource, as compared with nonrenewable fossil fuels, will be considered. Then, we will examine the potential for utilizing power generated at a central station for transportation. In the latter instance, an assumption will be made that the United States will become self-sufficient in energy when breeder reactors are producing a substantial fraction of our electric power.

The Hydrogen Fueled Automobile

Hydrogen, produced by the electrolysis of water, may be an excellent replacement for the world's dwindling automotive gasoline supply. Conventional internal-combustion engines have been operated on hydrogen fuel with almost no pollution and with increased efficiency. Engines designed specifically for hydrogen fuel can be expected to be even more efficient and pollution-free. In addition to transportation, the technological capability already exists to use hydrogen as a fuel in most other applications including space heating and cooling, and industrial processes. The biggest problem areas which remain appear to be associated with developing an overall H_2 production and distribution system which is safe enough for the average citizen to use. The space program provides a great deal of experience and technology which can be utilized toward the solution of these problems.

Consideration of hydrogen as a source of energy received its early impetus, in part, from the fact that it is a clean-burning fuel. Its heating value is 2.5 times larger than gasoline (62,050 Btu/lbm vs. 22,200) and the only by-products of combustion are water, if burned with pure oxygen, or water plus nitrous oxides, if burned with air. Even in the latter case, the amount of nitrous oxides is 5 to 14 times lower for hydrogen than for gasoline when burned in engines with the fuel mixtures ranging from lean to rich.

Hydrogen as a fuel for transportation can be stored in one of three ways, as a cryogenic liquid (LH_2), a compressed gas (GH_2), or as a metallic hydride. A comparison of fuel storage systems for a vehicle range of 260 miles is shown in Table 20.

Table 20. Comparison of Fuel Storage Systems for a Vehicle Range of 260 mi.[a]

	Gasoline	LH_2	GH_2	Metallic Hydride
Fuel:				
weight,lb	118	29.5	29.5	400 (MgH_2)
volume,ft³	2.6	6.7	35	8
	(20 gal)	(50 gal)		
Tankage:				
weight,lb	30	400	3000	100
volume,ft³	3	10	54	9
Total:				
weight,lb	148	430	3030	500

[a]After Stewart and Edeskuty.[5]

The weight of the LH_2 system reflects the fact that liquid hydrogen boils at 20°K and the tank is essentially a large Dewar requiring excellent thermal insulation characteristics. A compressed gas system is probably unreasonable in that a 3000-lb tank is required to store the hydrogen at 5000 psi. Certain metallic hydrides contain far more hydrogen per unit volume than does liquid hydrogen and a metallic hydride system is seen to be only 15% heavier than a corresponding LH_2 system.

Several properties related to the relative safety of gasoline and hydrogen are shown in Table 21.

Table 21. Comparison of Safety Parameters for Hydrogen and Gasoline.[a]

Property	H_2	Gasoline
Ignition energy (Btu)	1.9×10^{-6}	23.6×10^{-6}
Quenching distance (in)	0.024	0.098
Ignition temperature (°K)	1085	495
Combustion range (%)	4–75	1.5–7.6
Flame velocity (in/sec)	106	12

[a]From Stewart and Edeskuty.[5]

While a much higher temperature is required to ignite hydrogen, the energy required to ignite it is considerably lower. Thus, these factors tend to balance each other. Most of the other properties are pluses for gasoline. On the other hand, use of a metallic hydride for hydrogen storage is far safer than either gasoline or uncombined hydrogen, and it would eliminate a boil-off problem inherent in cryogenic storage systems.

Projections of the economics of a switch from gasoline to hydrogen for transportation fuel have been made.[5] The cost for a total H_2-system was found to be $133 billion. This should be compared to a National Petroleum Council forecast of $113-billion expenditure over the next 13 years for exploration, capital equipment, etc. for an expanded petroleum industry.

In summary, the technological capability for converting from gasoline to hydrogen for transportation energy exists. While the required capital investment for such a change is large, depletion of fossil fuels is imminent and a hydrogen based transportation system should be seriously considered.

The Electric Car

Advocates of electrically powered automobiles cite the arguments previously discussed as stimuli for accelerating development of such vehicles; namely, the fossil fuel and atmospheric pollution problems. The state-of-the-art of battery development was discussed in the preceding chapter. Here, we briefly look at the state-of-the-art of electric propulsion of automobiles and make some projections based on the assumption that the United States will become self-sufficient in terms of electric power in the not-too-distant future.

We have previously considered the current situation with respect to fossil fuel-fired internal combustion engines, and conjectured about transportation based on a hydrogen economy. In neither instance did we define explicitly the power and energy characteristics of a hypothetical standard automobile. Therefore let us, at this point, define some basic automotive terms and consider two standard vehicles, an urban car and a family car.

Torque is a measure of the ability of an engine to do work, while power is a measure of the rate at which the work is done. Understanding the difference between these terms will be particularly important when we compare electric and standard automobiles. Another way to put it is that torque determines whether an engine can drive a vehicle through snow, sand, or other obstacles, and the power determines how quickly the car passes over the obstacles.

The relationship between torque, T, and power, expressed as horsepower, hp, is

$$hp = \frac{TN}{5252}$$

where N is the engine speed in rpm. As an example consider a car moving at 60 mph. About 300 lb of force are needed to push the car along at this speed while overcoming the friction of the tires with the road and the aerodynamic drag on the body. The rate of doing work is then

$$300 \text{ lb} \times 88 \frac{ft}{sec} = 26,400 \frac{ft\text{-}lb}{sec}$$

or

$$\frac{26,400 \text{ ft-lb/sec}}{550 \dfrac{ft\text{-}lb/sec}{hp}} = 48 \text{ hp}$$

If the engine is turning at 3000 rpm, the amount of torque developed will be

$$\frac{26,400 \dfrac{ft\text{-}lb}{sec}}{50 \dfrac{rev}{sec} \times 2\pi \dfrac{rad}{rev}} = 84 \text{ lb-ft}$$

Figure 58 shows schematically the power output of an IC engine coupled to the drive wheels by a three-speed standard transmission and the output of a DC electric motor capable of driving the vehicle at the same top speed. The vehicle requirement curve represents all of the losses and drags including tire friction, aerodynamic drag, and internal losses of the engine and the drive train. At a given car speed, the vertical distance between the requirement curve and an output curve represents the reserve power available for accelerating and climbing grades. If this reserve is insufficient in a given gear, the driver may have to downshift. It should be noted that, at low speeds, an electric motor provides more capacity for acceleration.

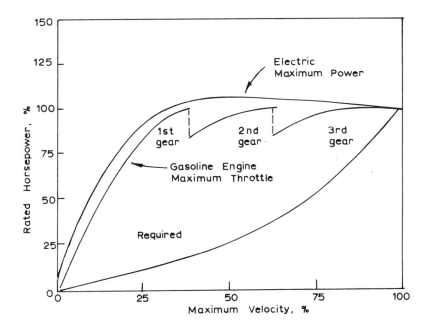

Figure 58. Typical power requirements.

An internal combustion engine at zero velocity is delivering zero torque and a mechanical friction clutch or fluid drive mechanism is needed to allow the engine to be "reved up" with respect to the wheel equivalent value. Figure 59 shows schematically the torque output curves for both the IC engine and a DC electric motor. It should be noted that the electric motor develops maximum torque at zero velocity. This characteristic is utilized by designers of so-called hybrid vehicles, which will be discussed in a subsequent section.

Assuming an electrically powered urban car of 2000-lb gross vehicle weight and a family car of 4000 lb, the energy density requirements for acceleration and cruising are given in Table 22.

Acceleration of the vehicle is dependent on the power density as shown, while the range of the vehicle is determined by the energy density. The problem facing us is getting enough power density and

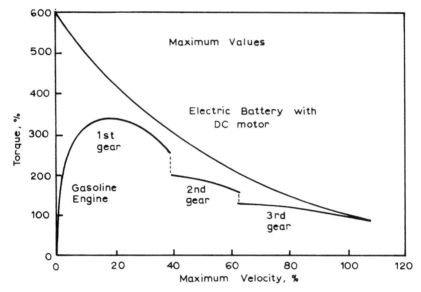

Figure 59. Typical torque characteristics.

Table 22. Energy Density Requirement for Acceleration and Cruise.[a]

| Type of Vehicle | Range (miles) | Constant Speed Cruise | | | Acceleration |
		Velocity (mph)	Energy Density (whr/lb)	Power Density (w/lb)	Power Density (w/lb)
Urban car	50	40	25	20	65
Family car	200	70	122	43	73–110

[a]After Hietbrink and Tricklebank.[6]

energy density in one battery system, without it being too heavy or too costly.

Power and energy requirements for various speeds and ranges for a 2000-lb vehicle are shown in Figure 60. The power and energy required to accelerate at less than half normal (for U.S. V-8s) and achieve a 200-mile range at 60 mph are, respectively, 80 hp and 50 kwh for an electric. The current state-of-the-art in batteries and fuel cells is shown in Figure 61. Areas representing internal combustion engines, as well as gas turbines and external combustion engines, are shown for comparison. Superimposed on the battery performance curves are the curves from Figure 60 representing the requirements for a 2000-lb vehicle cruising at 20, 40, and 60 mph over a range of 50–200 miles. Note that less than 3 hp is required to maintain steady-level driving at 20 mph while a 10-fold increase is required for 60 mph driving. The same is true of the energy required for driving range; 5 kwh will provide a range of 50 miles for steady driving at 20 mph, but 50 kwh is required for a 200-mile range at 60 mph.

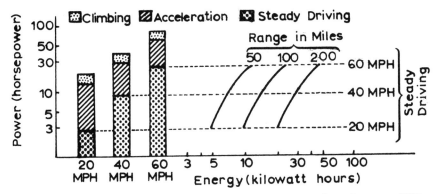

Figure 60. Power and energy requirements for various speeds and ranges (2000-lb vehicles). (Courtesy of *Electric Light and Power.*[7])

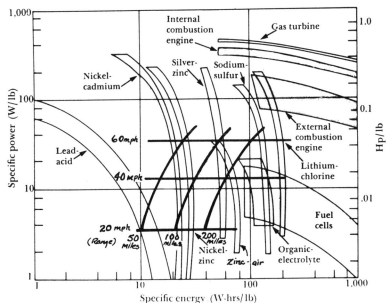

Figure 61. Vehicle requirements and motive power sources. When superimposed on current battery capabilities, no single 500-lb battery system would meet requirements of both range and acceleration. (assumes 2000-lb vehicle, 500-lb motive power source and steady driving. Power and energy taken at output of conversion device.) (Courtesy of *Electric Light and Power.*[7])

The data in Figure 61 show that conventional lead-acid batteries are not quite capable of supplying the energy and power requirements of the urban car given in Table 22. However, there is a substantial market for lead-acid batteries as a principal power source in transportation. In 1970, for example, approximately 750,000 in-plant trucks in the U.S., several thousand small delivery vans in England, and innumerable specialty vehicles such as golf carts used lead-acid batteries. This market is expected to double every ten years. The potential for increased application of electric batteries for transportation, particularly in the family car market, is contingent upon further development of inexpensive battery systems with high specific energy and power, such as the sodium-sulfur battery discussed in the preceding chapter.

The Hybrid Engine

A concept that appears to have considerable potential is the so-called hybrid vehicle. Such a vehicle is considered generally to be in the urban car class and it is powered by both a small heat engine and a battery system. In most designs, the battery system is utilized for acceleration because of the fact that it develops maximum torque at no load, while the heat engine supplies power for cruising. In addition, most hybrid designs allow for self-contained recharge of the batteries. This would be accomplished by having the heat engine drive a generator. A hybrid powered urban or commuter car would certainly relieve some of the demands made by transportation on our limited fossil fuels. However, since these fuels are limited, possibly the hybrid engine would serve best as a transition from heat engines to one of the other alternatives discussed.

SUMMARY

We have looked briefly at the problem of an increasing demand for transportation energy. In the long term, the realization that our high-energy density fossil fuels are limited and will eventually be depleted means that various transportation alternatives must be sought. Since the personal automobile is so firmly entrenched in our society, it seems highly likely that only those solutions which retain the personal automobile are viable. Two possible alternatives meeting this criterion were considered—utilization of hydrogen, a renewable fuel, and utilization of electrical power generated at a central station. The technology for instituting the first alternative is substantially existent. A major economic commitment would be necessary. The second alternative is contingent upon further development of battery technology, and the establishment of a self-sufficient electrical generating capability, whether through breeder nuclear reactors or large scale development of nuclear fusion or solar energy.

An additional source of energy for transportation should be mentioned. Converting electrical energy produced at a power plant to mechanical energy and storing it in a rotating, on board flywheel is under consideration. A considerable research and development effort toward optimum designs of flywheels and flywheel power drive mechanisms is underway both in the U.S. and elsewhere.

In the short- and intermediate-term, reduction of automobile weight, improvement in the thermal efficiency of heat engines, and encouraging more widespread use of transportation systems with higher propulsion efficiencies, are probably the most viable solutions to the transportation energy problem.

REFERENCES

1. Hirst, E. "Energy Consumption for Transportation in the U.S.," Oak Ridge National Laboratory, Report ORNL-NSF-EP-15, March 1972; *see also* Hirst, E., "How Much Overall Energy Does the Automobile Require?," *Automot. Eng.* 80, No. 7, 36-38 (July, 1972).
2. Goss, W. P. and J. G. McGowan. "Energy Requirements for Passenger Ground Transportation Systems," ASME Paper No. 73-ICT-24, Presented at the Second Intersociety Conference on Transportation, Denver, Colorado, September 23–27, 1973.
3. Harvey, D. G. and W. R. Menchen. *The Automobile, Energy, and the Environment* (Columbia, Maryland: Hittman Associates, Inc., 1974).
4. Hirst, E. "Pollution Control Energy Costs," ASME Paper No. 73-WA/Ener-7, Presented at the Winter Annual Meeting, Detroit, Michigan, November 11–15, 1973.
5. Stewart, W. F. and F. J. Eduskuty. "Alternate Fuels for Transportation. Part 2: Hydrogen for the Automobile," *Mech. Eng.* 22–28 (June, 1974).
6. Hietbrink, E. H. and S. B. Tricklebank. "Electric Storage Batteries for Vehicle Propulsion," ASME Paper No. 70-WA/Ener-7, presented at the Winter Annual Meeting, New York, 1970.
7. Marks, J. A. "Electric Vehicles: Big Market Without the Electric Car," *Electric Light Power,* 66 (January, 1970).

SUGGESTED FURTHER READING

1. Gratch, S. "Energy Consumption by the Transportation Industry," ASME Paper No. 73-WA/Ener-4, Presented at the Winter Annual Meeting, Detroit, Michigan, November 11–15, 1973.
2. *Intersociety Energy Conversion Engineering Conference Proceedings* (1968-1973).
3. Murray, R. G. and R. J. Schoeppel. "Emission and Performance Characteristics of an Air-Breathing Hydrogen-Fueled Internal

Combustion Engine," *1971 Intersociety Energy Conversion Engineering Conference*, Boston, Massachusetts (1971), pp. 47–51.
4. Rice, R. A. "System Energy as a Factor in Considering Future Transportation," ASME Paper No. 70-WA/Ener-8, Presented at the Winter Annual Meeting, New York, 1970.
5. Schurr, S. H. *Energy Research Needs* (Washington, D.C.: Resources for the Future, Inc., 1971), p. VII-20.

PROBLEMS

1. Write an equation for the straight line plotted in Figure 57.
2. a) Using data given in the Appendices, determine the number of Btus in one gallon of gasoline.
   b) Referring to Figure 57, determine the 1971 U.S. average mileage, (mpg of fuel consumed).
3. a) Replot the data given in Figure 57 as mileage vs. weight.
   b) Determine the constant for the equation
$$M = \text{constant}/w \text{ (mpg)}$$
   from the plot in part (a).
4. a) To a rough approximation, $M = 200/ D$, where D is engine displacement in cubic inches. Derive an expression between engine displacement and vehicle weight.
   b) List five factors other than vehicle weight that are also of importance in determining mileage.
5. To compare the overall energy efficiency of an electric automobile with one powered by a standard, Otto cycle (internal combustion) engine, one must account for the following efficiencies:

   e_t — thermal efficiency of the power plant
   e_{tr} — transmission efficiency
   e_c — efficiency of battery charging
   e_d — efficiency of battery discharging
   e_m — efficiency of electric motor.

   a) Estimate values for these various efficiencies.
   b) The overall efficiency of an electric auto will be the product of the above named efficiencies. Determine this based on your estimates in part (a) and compare to the efficiencies in Table 19.
6. A large-scale switch from gasoline powered to electric-powered automobiles would result in some changes in our overall pollution problems. Discuss in detail those changes.
7. Assuming the LH_2 tank for which data are presented in Table 20 in spherical and constructed from stainless steel ($\rho = 0.44$ lb/in³), determine the wall thickness required for a 50-gal tank. Neglect the insulation and vacuum space found in a typical Dewar construction.
8. In your opinion, would it be desirable to switch from gasoline to electric cars? Explain your answer in detail.

Chapter 8

Nuclear Fusion

INTRODUCTION

There are no commercial fusion power plants in the United States today and there are not expected to be any until the year 2000. Nuclear fusion is not a near-term solution to any of our current energy problems. The great attraction of nuclear fusion is this: if it could be made to work, many of our energy supply problems would be over. As we will see, there is enough nuclear energy in seawater to supply all conceivable energy needs for millions of years. As you may well imagine, however, the job of extracting this energy is not an easy one. There are no end of problems of fundamental understanding, as well as problems of engineering, which must be solved before the construction of commercial fusion power plants can even be considered. Indeed, fusion devices which can produce more power than they consume have not yet been built even on a laboratory scale, although such power-producing demonstration devices are expected to be achieved soon (perhaps this year). The time scale for the achievement of commercial nuclear fusion can possibly be estimated by remembering that the first sustained nuclear *fission* process was achieved by Enrico Fermi and coworkers at the University of Chicago on December 2, 1942. It took 15 years to produce the first commercial nuclear fission reactor, and even in the 1970s such fission reactors produced only a small percentage of our electric energy. Present fusion reactors are not yet even at a state equivalent to that of fission reactors in 1942.

FUSION PROCESSES

We have seen previously that certain very heavy elements such as $_{92}U^{235}$ may undergo fission into smaller nuclei when bombarded by neutrons. Since neutrons have no charge, they can readily penetrate into $_{92}U^{235}$ nuclei because there is no electrostatic repulsion between these two nuclear species. Energy is released in this reaction because

155

the mass of the resultant (fission) products does not equal that of the reactants (n plus $_{92}U^{235}$ nucleus). In nuclear fusion, two nuclear species are combined and, as in the fission process, energy is produced if the mass of the fusion products is less than that of the reactants. As an analogy to describe the fission process, one might consider the fusion of a neutron, $_0n^1$, and a hydrogen nucleus*, $_1H^1$,

$$_0n^1 + {_1H^1} = {_1D^2} \tag{58}$$

to make a deuterium nucleus, $_1D^2$. A deuterium nucleus is simply a hydrogen nucleus that contains one neutron in addition to a proton. Because there is no electrostatic repulsion, this reaction would occur readily *if the neutron interacted nonelastically with the hydrogen nucleus.* Instead, however, elastic interactions occur much more readily than nonelastic interactions. In such elastic processes, neutrons lose energy and are moderated, that is slowed down. This, of course, is why H_2O, among other substances, is used to alter the neutron flux in nuclear fission reactors. The exposure of $_1H^1$ atoms to a flux of neutrons is not, therefore, a viable way of producing energy.

Rather than the neutron-hydrogen fusion reaction, all fusion processes now considered as feasible for energy production involve the fusion of charged nuclear species. The species which can be made to undergo fusion most readily are deuterium, $_1D^2$, and tritium, $_1T^3$. A tritium nucleus is a hydrogen nucleus that contains two neutrons in addition to the proton. The fusion of $_1D^2$ and $_1T^3$ produces helium, $_2He^4$, and a single neutron:

$$_1D^2 + {_1T^3} \rightarrow {_2He^4} + {_0n^1} \tag{59}$$

A simple calculation shows that the weights of the reactants ($_1D^2$, $_1T^3$) and the products ($_2He^4$, $_0n^1$) are not equal.

Reactants	*Products*
$_1D^2$ ....3.3441 x 10^{-27}kg	$_2He^4$ ......6.6458 x 10^{-27}kg
$_1T^3$ ....5.0077 x 10^{-27}kg	$_0n^1$ ....1.6749 x 10^{-27}kg
Total .....8.3518 x 10^{-27}kg	Total .....8.3207 x 10^{-27}kg

The mass of the products is 0.0311 x 10^{-27} kg less than that of the reactants.

Mass data for other nuclides and atomic particles is given in Table 23. By means of the Einstein relationship

$$E = mc^2 \tag{60}$$

where E is the energy equivalent (in joules) of a mass m (kg) and

---

*A hydrogen nucleus is simply a hydrogen atom that has been ionized or stripped of its electron. As such it has a charge of +1.6 x 10^{-19} coulombs.

Table 23. Values for the Rest Mass (kg) of the first 11 Nuclides in the Periodic System plus Similar Data for Protons, Neutrons, and Electrons.

Element	Symbol	% of Natural Abundance (%)	Rest mass (kg)
Hydrogen	$_1H^1$	99.9851	1.6733 x 10^{-27}
(Deuterium)	$_1D^2$	0.0149	3.3441 x 10^{-27}
(Tritium)	$_1T^3$	—	5.0077 x 10^{-27}
Helium	$_2He^3$	10^{-4}	5.0077 x 10^{-27}
	$_2He^4$	100	6.6458 x 10^{-27}
	$_2He^5$	—	8.3222 x 10^{-27}
	$_2He^6$	—	9.9936 x 10^{-27}
Lithium	$_3Li^5$	—	8.3222 x 10^{-27}
	$_3Li^6$	7.5	9.9873 x 10^{-27}
	$_3Li^7$	92.5	11.6490 x 10^{-27}
	$_4Li^8$	—	13.3202 x 10^{-27}

Atomic Particles	Rest Mass (kg)
Electron	9.1096 x 10^{-31}kg
Neutron	1.6749 x 10^{-27}kg
Proton	1.6726 x 10^{-27}kg

c is the speed of light (2.299782 x 10^8 m/sec), the energy released by this fusion reaction can be calculated to be

$$E_{released} = (0.0311 \times 10^{-27}kg)(2.99792 \times 10^8 \frac{sec}{m})^2 \quad (61)$$
$$= 2.8131 \times 10^{-12} \text{ joules}$$

Since one electron-volt is equal to 1.6021 x 10^{-19} joules, this energy release may also be written as

$$\frac{2.8131 \times 10^{-12} \text{ joules}}{1.6021 \times 10^{-19} \frac{\text{electron-volt}}{\text{joule}}} = 1.7559 \times 10^7 \text{ eV} \quad (62)$$
$$= 17.6 \times 10^6 \text{ eV}$$
$$= 17.6 \text{ MeV}$$
$$= 2.82 \times 10^{-5} \text{ erg}$$

Thus, Equation (59) is usually written as

$$_1D^2 + _1T^3 \rightarrow _2He^4 + _0n^1 + 17.6 \text{ MeV} \quad (63)$$

This energy appears as the kinetic energy of both the helium and the neutron and is, in fact, not evenly divided between the two. The

$_2$He^4 particle is found to possess 3.5 MeV of kinetic energy while the neutron is found to possess 14.1 MeV of kinetic energy.

In a way, the relative division of energy is fortunate because, as we will see, several current fusion reactor designs rely on the high penetrating power of neutrons for the easy removal of this energy from the reactor. If most of the energy released were present in the $_2$He^4, totally different reactor designs would be needed.

CROSS SECTIONS

The probability that a $_1$D^2 nucleus and a $_1$T^3 nucleus will react according to Equation (59) is measured by their nuclear "cross section," σ(cm^2). Nuclear cross sections are not fixed, unique quantities for each type of particle but instead depend upon the exact conditions under which the reaction takes place as well as which specific reaction is being considered. For example, there will be one nuclear cross section for elastic interactions and quite another for nonelastic, fusion interactions.

To see how cross sections are measured and what they mean, imagine a beam containing N incident protons, that is N incident $_1$H^1 (hydrogen) nuclei, passing into a layer t(cm) thick containing Q of these $_1$H^1 nuclei *per square centimeter of layer area.* Suppose there then occurs n fusion reactions of the type

$$_1H^1 + {_1}H^1 = {_1}D^2 + {_1}e^0 \tag{64}$$

where, as before, $_1$D^2 represents a deuterium nucleus and $_1$e^0 represents a positron, that is, a particle having a mass equal to that of an electron but a positive rather than a negative charge. The "effective" volume swept out by the N incident $_1$H^1 particles passing through the layer t is σNt cm^3, while the average volume per $_1$H^1 particle in the layer is

$$\text{Average Volume per } {_1}H^1 \text{ particle} = \frac{t \text{ cm}}{Q\left(\frac{1}{\text{cm}^2}\right)} = \frac{t}{Q} \text{ (cm}^3) \tag{65}$$

The number of fusion collisions, n, is then given by

$$n = \frac{N\sigma t}{\frac{t}{Q}} = \sigma NQ \tag{66}$$

or

$$\sigma = \frac{n}{NQ} \text{ (cm}^2) \tag{67}$$

The cross sections σ for nuclear processes can vary widely (*e.g.*, from 10^{-20} cm^2 to 10^{-32} cm^2. For convenience, a nuclear cross section of 10^{-24} cm^2 is called a "barn." In understanding what a cross section represents, it is important to repeat that it does not represent an actual physical cross section. Rather, nuclear cross sections, in effect, are simply a convenient bookkeeping way of keeping track of the probability that two particles will interact in a certain way. Look again at Equation (66) and note that a number of reactions (in this case fusion collisions) will clearly increase as either the number of incident particles, N, or the area density Q, of particles increases. The cross section, σ, is simply the constant of proportionality which tells how fast the number of fusion reactions will increase as either N or Q increases.

There is no single nuclear cross section for a given particle. Instead, nuclear cross sections will depend on a) the type of nuclear particles that are interacting, b) the energy with which they are being forced together, and c) what type of interaction is being considered. As we will see, the concept of nuclear cross sections is very useful in deciding which nuclear processes may be considered for practical fusion energy systems.

THERMONUCLEAR FUSION

Nuclear species having the same electric charge, such as $_1D^2$ and $_1T^3$ nuclei, will fuse together according to Equation 59 or Equation 63 only if they are driven together with sufficient energy to overcome their electrostatic repulsion. This energy could be supplied by accelerating the charged particles with electrostatic fields. Indeed, accelerators have been and are being used to determine nuclear cross section data for all fusion reactions. The use of accelerated particles to generate power is not practical only because of the inefficiency of present acceleration methods. Vast amounts of energy are needed to produce beams whose density is far too low to give significant amounts of fusion energy. A more practical procedure is to use the thermal energy given to the particles as the temperature is raised. Nuclear reactions brought about by an increase in temperature are called thermonuclear reactions or, in this case, thermonuclear fusion.

The average kinetic energy of a particle, whether an electron or a nucleus, at a temperature T is given by

$$\text{Average Kinetic Energy} = 3/2 \, KT$$

where K is Boltzman's constant (1.38×10^{-16} erg/°K) and T is the temperature in degrees Kelvin (°K). One electron-volt of energy thus corresponds to 7.7×10^3 °K. Fusion reactions proceed at appreciable rates only when the fusion particles having energies above a few tens of KeV (tens of 10^3 eV). Such energies correspond to tens of millions of °K. At such temperatures all substances are fully ionized, that is, the nuclei are completely stripped of their surrounding electron clouds. These fusion reactions generally proceed at an increasing rate as the temperature is raised because the reaction cross sections increase. Figure 62 shows, for example, nuclear cross section data for four different fusion reactions as a function of temperature. The reaction involving deuterium and tritium (the D-T reaction) has been discussed already. The other three reactions are

$$\text{(D-D)} \quad {}_1D^2 = {}_1D^2 \Big\langle \begin{array}{l} {}_2He^3 + {}_0n^1 + 3.2 \text{ MeV} \quad \text{(68a)} \\ {}_1T^3 + {}_1H^1 + 4.0 \text{ MeV} \quad \text{(68b)} \end{array}$$

$$\text{(T-T)} \quad {}_1T^3 + {}_1T^3 \rightarrow {}_2He^4 + {}_0n^1 + {}_0n^1 + 11.33 \text{ MeV} \quad \text{(69)}$$

$$\text{(D-He}^3) \quad {}_1D^2 + {}_2He^3 \rightarrow {}_2He^4 + {}_1H^1 + 18.3 \text{ MeV} \quad \text{(70)}$$

As may be readily seen from Figure 62 the D-T reaction shows much higher cross sections at much lower temperatures than any of the other reactions. Consequently, thermonuclear fusion system designs have been primarily based on the D-T reaction. However, it is worth noting that in the D-He³ reaction, 100% of the energy that is released is carried by charged particles. As we have seen previously, for example, in the D-T reaction only 3.5/17.6 = 19.9% of the energy that is released is carried by the charged ${}_2He^4$ ions. In some thermonuclear designs, it is proposed that electric power be produced directly from the manipulation of charged particles in magnetic fields, without the use of stream or other heat-to-electric power conversion cycles. For such designs the D-He³ reaction is favored even though, as may be seen from Figure 62; high plasma temperatures are required.

It should be noted that all of the fusion reactions that have been suggested involve deuterium. This situation is fortunate because, of the three nuclides ${}_1D^2$, ${}_1T^3$, He^3, only ${}_1D^2$ occurs naturally to any sensible degree (Table 23). As atomic, rather than nuclear species, both deuterium and tritium can combine with oxygen to form "heavy" water and both D_2O and T_2O are more dense than normal H_2O. Only D_2O occurs naturally, however, and eight gallons of ordinary water contain about 1 gram of deuterium. The deuterium may be extracted from the normal, light H_2O, by distillation, since D_2O and H_2O have different vapor pressures. The cost of this extraction is only about 30 cents per gram of deuterium. If this one gram of deuterium were combined according to Equation 68 (both possible reaction product sequences occurring with essentially the same probability) the energy produced would be equal to about 2500 gallons of gasoline. There are vast quantities of deuterium available from the enormous quantity of water present in the oceans (approximately 10^{13} tons of deuterium). This vast deuterium resource is the basis of the oft-heard statement that fusion is capable of supplying all our energy needs for eons to come. Even at 1000 times the present rate

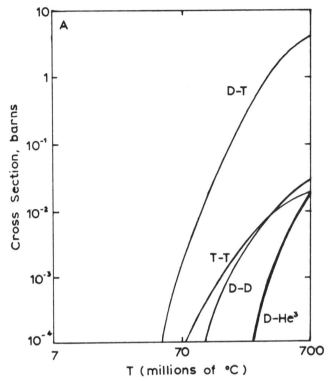

Figure 62. Nuclear cross sections versus plasma temperature for four different thermonuclear fusion reactions. (Courtesy of the Institute of Electrical and Electronics Engineers.)

of world consumption, there would be more than enough energy to last one billion years.

To obtain this energy by thermonuclear methods, it is not enough that fusion merely occur. Rather, fusion energy must be produced at a rate which is greater than the rate at which energy is lost from the plasma within which fusion is occurring. In thermonuclear systems, the principal cause of such energy loss which must be considered is *Bremsstrahlung* or "breaking radiation." This radiation, in the form of gamma rays, inevitably occurs during the deceleration of the charged particles which compose the plasma when these particles collide. The resultant energy loss for a given plasma is linearly proportional to the plasma density and the temperature. As seen in Figure 62 the fusion cross section increases much faster than linearly with temperature. Therefore, at some temperature, the fusion process will produce more energy than that lost in *Bremsstrahlung,* as shown in Figure 63. This temperature is called the critical ignition temperature and is reached, as an approximation, when the reaction cross section reaches about 5 x 10⁻³ barns. From Figure 62, the ignition temperature for the D-D fusion reaction is about 500,000,000°K, while that for the D-T fusion reaction is only about 100,000,000°K. It is for this reason that current interest is focused on the D-T fusion reaction even though tritium does not occur naturally in significant quantity and must itself be produced by a nuclear reaction. As we will see in the next section, tritium production is an integral part of most fusion reactor designs.

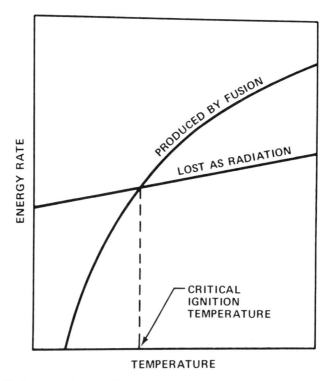

TEMPERATURE

Figure 63. A comparison of the rate of energy production produced by fusion versus the rate of energy loss by radiation, both as a function of plasma temperature.

In addition to achieving the ignition temperature, any thermonuclear fusion reactor must achieve a practical absolute rate of power generation. A useful figure of merit for considering such power generation is the product of the plasma density, n (particles per cubic centimeter) and the time, τ, (sec) during which the plasma is at that density. For a practical fusion device using the D-D reaction this product must be at least equal to 10¹⁶ sec/cm³, but only 10¹⁴ sec/

cm³ for the D-T reaction. These values are known as Lawson criterion and are another reason why the D-T reaction appears to be the most practical fusion reaction. To achieve these values, any practical combination of confinement time and plasma density may be used. For example, a D-T plasma having a density of 10^{14} particles per cubic centimeter would need to be contained for one second or more while a more dense plasma of 10^{16} particles per cubic centimeter would need only 0.01 seconds to achieve the same effect. As we shall see, the achievement of dense, stable plasmas at temperatures approaching 100,000,000°K has proven to be an elusive, difficult, engineering feat.

With this review of the fusion process, we go on now to briefly outline some of the fusion reactor schemes which have been proposed.

PROPOSED ENGINEERING SCHEMES

At temperatures above 20,000–30,000°K, all substances are completely ionized. That is, the atoms are stripped of their surrounding electrons and only positively charged nuclei remain. Thus, at anything approaching fusion ignition temperatures, one is no longer dealing with solids, liquids or gases, but rather with a mixture of free, negatively charged electrons and free positively-charged nuclei, a plasma. The principal problem to be overcome is the containment of this plasma, and its compression to achieve a high density, n, for a long period of time, τ.

Any interaction of the plasma with the walls of the fusion reactor would quickly drop the temperature of this plasma to below the ignition point. Because the plasma is composed of charged particles, however, it will interact with any exposed external magnetic or electric fields. The containment of the plasma by means for example, of electrified spherical containment vessels is not possible because the net field on the inside of a sphere that is charged to any given potential is identically zero. Because of this fact, the electric field from the sphere would have no effect at all on the charges inside. Primarily, one is left with only magnetic fields or inertial effects.* The interaction between charged particles and magnetic fields is as follows: if the charged particle is not moving or is moving parallel to the magnetic field lines, there is no interaction. If, on the other hand, the charged particles are moving perpendicularly to the magnetic field lines, there will be a force on each particle and this force will be perpendicular to both the magnetic field and the direction of

---

*In D-T hydrogen bombs the particles remain together long enough (10^{-7} sec) for fusion by inertia alone. A fission bomb is used as the "match" to ignite the mixture.

motion of the particle. The magnitude of this force is proportional to the charge on the particle, its velocity, and the magnetic field strength. Because of this force, moving charged particles either circle or spiral around the magnetic field lines. If there is no velocity component parallel to the magnetic field lines the orbit of the particle is circular; if the particle does have some velocity parallel to the magnetic field, then it moves in a spiral. Similar spiralling accounts for the fact that the Aurora Borealis and Aurora Australis increase in intensity as one goes toward the north and south poles (where magnetic field lines enter or leave the earth).

Fundamentally, these different magnetic containment schemes can be broken down into two different types: open-ended systems and closed systems. Open-ended systems, like that shown in Figure 64, prevent plasma from escaping through the cylinder walls because, as we have seen, charged particles moving perpendicularly to field lines tend to spiral around their field lines. Escape of the plasma through the ends of the cylinder is minimized by making the magnetic field locally stronger at these end regions, as shown. This increase in field strength produces a "magnetic mirror" at the cylinder ends and, particles will tend to be "reflected" back into the interior of the cylinder. Particles, which are moving along the axis of the cylinder will still escape, however, since they are moving parallel to the magnetic field lines. Thus, open-ended systems tend to be "leaky," as one might expect. Advantage can be taken, however, of such leaks to produce D.C. electric power directly; as shown in Figure 65, the plasma leaking through the ends of the system can be used directly to produce electric power. The electrons, being thousands of times lighter than the ions are deviated more strongly by the weakening magnetic field and hence are separated from the positive ions. In such a system, only the energy which is deposited in charged particles can be used to generate electric power. In a system using the $D\text{-}_2He^3$ reaction, 100% of the energy produced by fusion appears in such particles (Equation 70), and this fact accounts for the particular interest in this reaction, even though, as shown in Figure 62, it would involve much higher plasma temperatures. In such a direct conversion process, very high efficiencies, approaching 90%, can be obtained. This result decreases the Lawson criterion values, which are based on 33% efficient energy conversion to find the $n\tau$ product required for a practical fusion system. Other open-ended systems have involved the rapid compression or "pinching" of plasma by the rapid buildup of magnetic field intensity by means of enormous currents (millions of amperes) by rapidly discharging capacitor banks through single turn coils as shown in Figure 66. Since the current flow is in the so-called theta direction, such devices are called theta pinch machines. Such systems have produced temperatures of

50,000,000°K and plasma densities of 5 x 10^{16} particles per cubic centimeter. Because of the rapid escape of the plasma, the Lawson criterion are missed by a factor of 1000, due to the extremely short containment times. Systems combining magnetic mirrors with the theta pinch design have been built in an attempt to increase these containment times.

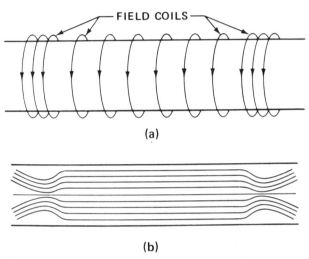

(a)

(b)

Figure 64. Schematic drawing of an open-ended magnetic mirror confinement system showing (a) disposition of magnetic field coils and (b) the resultant magnetic field lines.

To increase containment time and decrease plasma leakage, it is also possible to construct closed magnetic containment systems. These systems are most often in the shape of a doughnut or toroid. One of the best-known of these close-ended designs is the so-called Tokamak arrangement, which was first proposed at the Kurchatov Institute (Moscow) and reported in 1959. This design is shown schematically in Figure 67. The current in the field coils around the toroid generates a magnetic field which passes around the doughnut axis of the toroid in just the same way as the coils around a cylinder produce a field parallel to the axis of the cylinder (Figure 64). Any plasma inside the toroid will be initially contained by this magnetic field as the charged plasma ions spiral around the field lines. The transformer arrangement shown passing through the toroid may be used to produce the initial plasma. That is, by passing a very large current pulse through the primary windings, a current is induced in the interior of the toroid (Figure 68), which then in effect, acts as a simple secondary winding of the transformer. The current carried by this plasma itself produces a magnetic field (Figure 69) normal

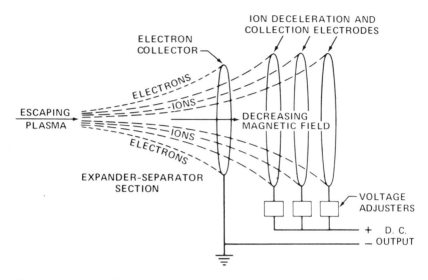

Figure 65. A possible scheme for the direct conversion of plasma energy to electric energy by means of the separation of the electric charges in the plasma.

to the doughnut axis of the toroid. This field is called the poloidal field. The toroidal and poloidal fields interact to produce a resultant spiral magnetic field encircling the plasma.

It will be remembered that, although $_1D^2$ is a naturally occurring isotope, $_1T^3$ is not. Thus, in Tokamak systems designed around the

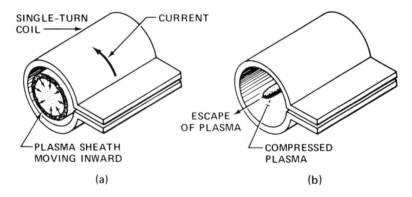

Figure 66. Schematic drawing of a theta pinch device showing (a) initial plasma compression by a large current pulse followed by (b) quiescent compressed plasma escaping through the open system ends.

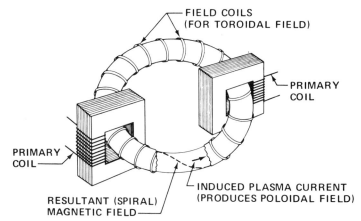

Figure 67. The basic components of Tokamak systems. The primary coils are used to produce the current in the plasma, as shown in Figure 68.

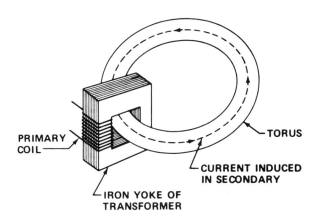

Figure 68. The plasma acts as a single turn secondary coil of the transformer.

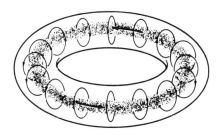

Figure 69. The current in the plasma produces a poloidal field which combines with the initial toroidal field as shown in Figure 67.

D-T reaction, provision must be made to produce $_1T^3$ by means of the neutrons produced by this reaction.

To generate $_1T^3$, the entire toroid can be surrounded with liquid lithium which would serve a dual purpose—tritium generation and heat extraction. As we have seen, most of the fusion energy produced in the D-T reaction resides in the neutron. The energy in this neutron can be extracted and $_1T^3$ produced by the interaction of these neutrons with lithium by either elastic or fusion collisions. The way in which lithium fuses with these neutrons depends upon whether the neutrons are fast (high energy) or slow (low energy) neutrons. In any case, however, tritium is produced:

$$_3Li^7 + \text{fast neutron} \rightarrow {}_1T^3 + {}_2He^4 \text{ slow neutron} \qquad (71)$$

$$_3Li^6 + \text{slow neutron} \rightarrow {}_1T^3 + {}_2He^4 \qquad (72)$$

Whether the incident neutrons fuse with or interact elastically with the lithium, substantially all the energy of the incident neutron is transferred to thermal energy. The general idea then is to make the chamber wall of a refractory metal, *e.g.*, niobium which is relatively transparent to 14.1 MeV neutrons, and then to surround this wall with a layer of liquid lithium. A suggested lithium layer thickness in a full-scale reactor is one meter. The lithium must be liquid in order that heat and the tritium may be extracted from it. It is this extracted heat which is used to produce usable electric power, via tried and true steam turbines.

There are many engineering problems associated with this design, as well as other magnetic containment designs, and we will mention some of these problems below. The Tokamak arrangement, however, has already been able, at least in the Russian device, to produce confinement times and plasma densities that are only about one order of magnitude below the Lawson criterion for the D-T reaction. Presently in the U.S., there are seven major laboratories engaged in closed-system fusion research programs.

There are naturally a great many difficulties associated with the Tokamak design as well, of course, as with all other magnetic containment designs. For example, the vast neutron flux which would pass through the inner niobium wall would give rise to tremendous radiation damage. There is even doubt that niobium, or any other material, could long stang up to this high neutron flux. Secondly, in any viable commercial fusion reactor, it will be necessary to use superconducting coils to generate the magnetic field used to contain the plasma. A normal conductor, *e.g.*, copper, dissipates energy, in the form of heat, equal to the square of the current multiplied by the resistance of the copper. A superconductor has no resistance to the flow of current and therefore does not dissipate any energy during operation. Unfortunately high magnetic fields can destroy this superconductivity. It will therefore be necessary to develop superconducting materials which can withstand very high magnetic fields.

Finally, a very critical problem involves the stability of the plasma cloud being confined by the magnetic field. This problem is illustrated by the results shown in Figure 70. The basic problem is this: the plasma consists of charged nuclei and electrons. Due to their motion, these particles themselves produce magnetic fields. The particles are thus affected not only by the external magnetic fields but also by the magnetic and electric fields produced by the plasma itself. As shown in Figure 70, the development of these instabilities generally means that the plasma will touch the walls of the chamber and rapidly cool. For open-ended systems, the problem of plasma instabilities is considered generally less important than the question of plasma leakage. For closed systems, however, such as those based on toroidal chambers, the question of plasma instabilities and the concomitant inability to achieve plasma confinement for longer times is still of critical importance. The whole question of plasma stability is exceedingly complex and would seem a fruitful ground for detailed mathematical analysis and computer simulation.

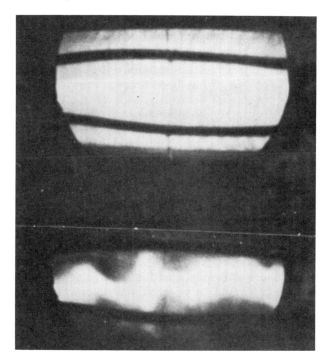

Figure 70. A photograph showing the development of instabilities in a plasma a few microseconds after formation. (Courtesy of Los Alamos Scientific Laboratory.)

Finally, combined fusion-fission designs have been suggested in which the lithium blanket contains also $_{92}U^{238}$. The idea here is the use of the high neutron flux to breed plutonium from the $_{92}U^{238}$. Preliminary calculations seem to show that such systems might have better prospects of producing electric power economically than pure fusion reactors because of the value of the plutonium that would be produced. Naturally, such new designs also involve a host of additional engineering problems.

LASER-IGNITED FUSION

A completely different approach towards igniting the D-T fusion reaction was started in the 1960s. In this approach laser heating is combined with the inertial confinement already mentioned in connection with thermonuclear weapons. A well-focused pulse of laser light is used to attempt to very rapidly heat a small pellet of deuterium and tritium to temperatures on the order of 100 million °K. Newton's law of inertia largely determines the speed of expansion of the pellet. If several lasers are used simultaneously, the force of the expanding outer pellet layers compresses the remaining pellet inner layers, thus increasing the rate of the fusion reaction. By shaping the intensity of the pulse in time the pellet can be compressed more effectively than otherwise. Currently, however, the energy produced by the resulting fusion is several orders of magnitude less than the energy delivered by the lasers. As an example of the engineering complexity of this approach, it has been proposed that it may be necessary to hit the pellet simultaneously with 27 of the highest power neodymium lasers currently available in order to approach the break-even point (as much fusion power out as laser power in). As in the Tokamak design, the fusion chamber should be surrounded by lithium both to remove heat and to generate tritium. Among the other special problems of the laser-fusion approach is the need for a containing wall strong enough to contain the fusion blast. For a break-even fusion reactor, this blast is estimated to be equivalent of 500 pounds of TNT. Figure 71 illustrates schematically how a laser fusion system would operate. A deuterium-tritium pellet is first injected along the vortex of a swirling lithium film flowing across the interior of a spherical reactor vessel. When the pellet reaches the center, a pulsed (10^{-9}sec) laser beam or beams would ignite this pellet. After a short wait, another pellet would be injected, and so on. The neutrons produced in the resultant D-T reactions would then be taken up by the lithium film where they would deposit their kinetic energy as well as breed tritium. This lithium would then pass through a tritium removal stage and heat exchangers for steam generation before being returned to the reactor. Clearly, the engineering problems for this

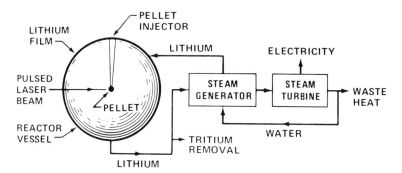

Figure 71. Conceptual arrangement of a laser-initiated fusion system.

reactor are substantially different from those for ones based on the Tokamak design.

Laser systems have been much less studied than magnetic containment systems. As an example of the state of development of laser systems, there is considerable difference of opinion regarding the optimum size of the deuterium-tritium pellet. Nevertheless, laser fusion is considered a very promising approach. One very great advantage of the laser approach is that commercial power production might be achievable with reactors as small as a few hundred megawatts. Tokamak systems would need to be much larger, on the order of 5000 Mw.

SAFETY CONSIDERATIONS

Although it is universally agreed that fusion reactors, under no conceivable circumstances, can turn into hydrogen bombs, there are other dangers, Chief among these problems is that of tritium. Tritium is a radioactive isotope and decays according to

$$_1T^3 \rightarrow {}_{-1}e^0 + {}_2He^3 \tag{73}$$

with a half-life of 12.1 years. The symbol $_{-1}e^0$ represents an electron.

It is estimated that a 5000-Mw Tokamak fusion reactor would, all told, contain 2×10^8 curies of tritium. A curie of radioactive material is that amount which will have 36.4×10^9 disintegrations per second. As an example of how large a curie is, it is worth noting that it is usual to require AEC registration of amounts of radioactive materials on the order of millicuries. Furthermore, 2×10^8 curies of tritium are essentially equivalent to the amount of the most hazardous fission product (10^8 curies of $_{53}I^{131}$) to be found in a full-scale breeder reactor. Tritium itself, as the gas T_2 (equivalent to H_2) is relatively benign since it is not absorbed into the body and would float up into

172   *Introduction to Energy Technology*

the outer reaches of the atmosphere. As T_2O, however, tritium would enter into the life cycle in the same way as water. Furthermore, T like H can diffuse through substances more rapidly than can any other element. Containment of such vast quantities of tritium has never before been attempted and can therefore be expected to present substantial new engineering problems. In the worst case, where all the $_1T^3$ escaped as T_2O, these 2×10^8 curies would be enough to render undrinkable 20 trillion gallons of normal water. If a way could be found to "burn" pure deuterium (the D-D reaction) rather than deuterium-tritium mixtures (the D-T reaction) the quantities of tritium involved would decrease by many orders of magnitude. This decrease would occur, of course, primarily because the need to generate tritium using lithium would be eliminated. It is probably a moot question at this point whether the higher ignition temperature and higher Lawson criterion for the D-D reaction presents a worse problem than merely dealing with the vast quantities of tritium involved with a fusion reactor using the D-T reaction. There are clearly many problems to be solved in either case, but the pay-off is so tremendous that there is no doubt that research on a large scale will continue.

SUGGESTED FURTHER READING

1. Glasstone, S. *Controlled Nuclear Fusion,* U.S. Energy Research and Development Administration, Office of Information Services.
2. Robbins, L. W., (ed.) *Survey of the USAFC Program in Controlled Thermonuclear Research,* USAFC, Division of Thermonuclear Research, Washington, D.C.
3. Rose, P. J. "Controlled Nuclear Fusion Status and Outlook," *Science* 172, 797 (1971).
4. Metz, W. D. "Laser Fusions: A New Approach To Thermonuclear Power," *Science* 177, 1180 (1972).
5. *Survey of Fusion Reactor Technology,* Report EUR 4873 e (Washington, D.C.: Euratom Fusion Reactor Technology Advisory Group, European Community Information Service, 1972).
6. Post, R. F. "Nuclear Fusion by Magnetic Confinement," *1973 IEEE Intercon Energy Utilization and Control* (New York, N. Y.: Institute of Electrical and Electronic Engineers, 1974), p. 6/1.
7. Boyer, K. "Laser Initiated Fusion — Its Problems and Promises," *1973 IEEE Intercon Energy Utilization and Control* (New York, N. Y.: Institute of Electrical and Electronic Engineers, 1974), p. 6/2.
8. Bishop, A. S. *Project Sherwood The U.S. Program in Controlled Fusion* (Reading, Massachusetts: Addison-Wesley Publishing Co., 1958).
9. Green, T. S. *Thermonuclear Power* (New York, N. Y.: Philosophical Library, 1963).

10. Glasstone, S. and R. H. Lovberg. *Controlled Thermonuclear Reactions* (New York, N. Y.: Van Nostrand Reinhold, 1960).
11. Gough, W. C. and B. J. Eastland. "The Prospects of Fusion Power," *Scient. Amer.* **224**, 50 (February, 1971).
12. Post, R. F. "Prospects for Fusion Power," *Physics Today* **26**, 30 (April, 1973).

PROBLEMS

1. The sun delivers 1.36 kilowatts per meter squared onto the earth. The radius of the earth's orbit is 93,000,000 miles.
   (a) At what rate, in kilowatts, is the sun producing energy?
   (b) At what rate, in kilograms per second, is the sun converting matter into power?

2. The thermonuclear reactions assumed to occur in the sun involve the conversion of four $_1H^1$ nuclei into one $_2He^4$ nucleus. Calculate the energy released, in kilowatt hours, per gram at $_1H^1$ that reacts in this way.

3. A theta pinch device is designed to heat 10^{21} $_1D^2$ atoms and $10^{21}$$_1T^3$ atoms by the rapid discharge of 300,000 joules of energy from a capacitor bank. What plasma temperature will be reached if 25% of this energy is deposited within the plasma and this energy is equally shared by the electrons and the nuclei?

4. Suppose a beam containing a total of 10^{16} deuterium ions is passed through a tritium ion cloud which contains 10^{20} $_1T^3$ ions and which is 10 cm x 10 cm x 2 cm thick. If the velocity of the incident deuterium ions is equivalent to that which would be obtained at a temperature of 700,000,000°K, how much fusion energy in kilowatt hours, would be produced?

$$1 \text{ kwh} = 3.6 \times 10^{13} \text{ erg} \quad 1 eV = 1.6 \times 10^{-12} \text{ erg}$$

5. Suppose a beam of 10^{12} $_1H^1$ ions is passed through a layer of $_1H^1$ ions which has an area density of 10^{19} $_1H^1$ ions per square centimeter. It is found that a total of 10^2 positrons are produced. What is the cross section in barns, for the H-H reaction under these conditions?

6. Nuclear particles can be given extremely high energies by accelerating them using both electric and magnetic fields. These high energies could be reached only by plasmas having truly incredible temperatures. Calculate the plasma temperature which would have to be reached in order to produce $_1H^1$ particles having average energies equivalent to those produced by a typical 1,000,000,000-volt accelerator. Hint: 1 electron moving through 1 volt acquires 1 eV of energy.

7. Discuss the probable impact on underdeveloped countries if the United States successfully developed large-scale fusion power generation.

8. Discuss the probable impact on the U.S. if the Soviet Union successfully developed large-scale fusion power generation and we did not.

9. Discuss briefly the principal problems to be overcome in Tokamak thermonuclear systems versus those to be overcome in laser-ignited fusion systems.

10. Will fusion reactors supply 5% of our electric power requirements by the year 2015? Justify your answers.

Chapter 9

Solar Energy

INTRODUCTION

Solar energy is a renewable resource; it cannot be depleted. The sun constantly delivers 1.36 kw (1360 joules/sec) of power per square meter (430 Btu/hr-ft^2 or 123 w/ft^2) to the earth. Naturally, some of this power is absorbed by the atmosphere so that even at 12 noon on a sunny day in desert areas of the southwest the land surface may only receive about 1000 w/m^2 or 0.1 w/cm^2. This reduction in solar energy power by the atmosphere is shown in Figure 72. Note in this figure that the units of the ordinate are (w/m^2)/cm^{-1}. The units must be this way because the data show the power delivered as a function of the wave number (cm^{-1}) of the light received. The wave number, v, can be converted to the wavelength, λ, since $v = 1/\lambda$. The measured intensity, w/m^2, will depend on what range of the spectrum is included. This effect is accounted for by reporting the intensity as (w/m^2)/cm^{-1}. As the curve shows, the atmosphere absorbs particularly strongly at several wave numbers (or wavelengths). This selective absorption is due primarily to the presence of CO_2, -H_2O, and O_3 in the atmosphere. It is worth noting that the very short ultraviolet portion is virtually eliminated by atmospheric absorption. It is the integral of the resultant intensity over all wave numbers which yields the value of approximately 1000 w/m^2. The intensity of sunlight also, of course, varies with the season, as is shown for three locations (in California) in Figure 73. It is also obvious, of course, that the sun only shines for a limited amount of time each 24 hours. Thus, the 1 kw/m^2 intensity of sunlight, must be decreased substantially if one is considering the average over 24 hours. For regions of the southwest the average over 24 hours of sunlight intensity is typically about 260 w/m^2. Different values would be found in other locations, and Figure 74 shows the combined influence of all factors, including weather, on the solar power, in Btu/

ft² — average day, across the continental United States. As may be seen, there is a variation of a factor of two between the deserts of the southwest and the north-central regions.

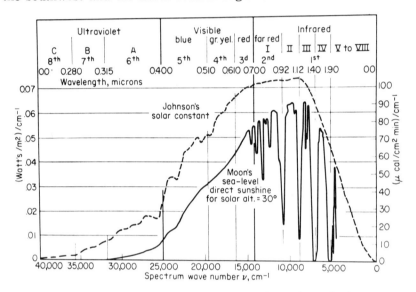

Figure 72. The effect of the atmosphere on the incident solar radiation spectrum (Note: "Moon's" refers to the investigator, not to our satellite (From Zarem and Erdway, *Introduction to the Utilization of Solar Energy,* Copyright 1972, McGraw-Hill Book Co.)

In spite of these effects, the *total* current electric generating capacity of the United States for 1974 (about 0.44 trillion watts), could be equaled by collecting (and converting to electricity with 10% efficiency) the maximum solar energy which falls on a square area of land, in the southwest, not more than 40 miles on a side. Even considering that the sun does not shine all day only requires a factor of four increase in area, to a square 80 miles on a side to produce 0.44 million watts averaged over 24 hours. The reader should check the calculation of these numbers; many reports have come up with far larger area requirements than appear to actually be needed. Much of the southwest is sparsely populated desert, and there would seem to be plenty of room for this size installation. In addition, it may be possible to use large numbers of smaller electric generating units, suitable in size for individual homes, rather than a few large installations. In any case, the potential value of solar energy is enormous. The principal problem to be overcome is the development of economically viable engineering ways to collect and use this vast, inexhaustible energy resource. In what follows we will review some of the engineering methods that have been proposed or already at-

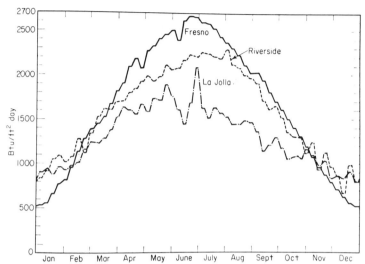

Figure 73. The annual variation in solar power for three different locations in California. (From Zarem and Erdway, *Introduction to the Utilization of Solar Energy,* Copyright 1972, McGraw-Hill Book Co.)

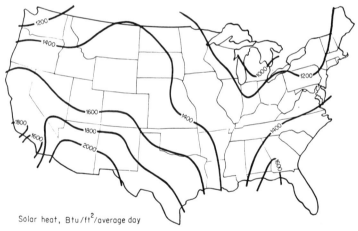

Solar heat, Btu/ft^2/average day

Figure 74. The distribution of average daily solar energy in the United States. (Courtesy of Industrial Press, Inc.)

tempted, with emphasis on the engineering problems which remain to be solved before the promise of solar energy can be realized.

THERMAL SYSTEMS AND THE SOLAR ENERGY COLLECTION PROBLEM

A major problem in the practical utilization of solar energy is the initial collection of this energy. We already have seen in earlier chapters, that the efficiency, η, of the best possible (Carnot) heat engine depends upon the high, T_H, and the low T_L, temperatures between which this engine operates:

$$\eta = \frac{T_H - T_L}{T_H} \tag{74}$$

It is evident, therefore, that heat engines that might operate using solar energy will be more efficient the higher the temperature source provided by the collector of this solar energy. To estimate the temperatures that can be reached through the collection of solar energy, remember that a black piece of rubber hose placed in sunlight will heat the water inside this hose to a higher temperature than will a similar white piece of hose. Clearly, the steady-state temperature that the hose will reach, assuming no wind or other air convection and no water flowing, will be that at which the heat lost from the hose by radiation equals the heat supplied to the hose by the sunshine. To quantify this conclusion so that an actual temperature can be calculated, some basic laws of radiation heat transfer must be invoked. For our purpose, the most important of these laws is that which was deduced independently by Joseph Stefan (1835–1890) and Ludwig Boltzmann (1847–1906). The latter is the same Boltzmann for whom the constant is named. Stefan deduced this law by experiment. Boltzmann deduced it by pure thermodynamic reasoning. They both came to the same conclusion; namely, that the power per unit area, $(P(w/cm^2)$, radiated by any body at a temperature $T(°K)$ is proportional to the fourth power of that temperature. The constant of proportionality in this result will clearly depend on the substance being kept at the temperature T and to take account of this, the Stefan-Boltzmann law is written as

$$P \frac{w}{cm^2} = \sigma e\ T^4 \tag{75}$$

where σ is the Stefan-Boltzmann constant $(\sigma = 5.67 \times 10^{-12} \frac{watts}{cm^2 °K^{+4}})$ and e is the emissivity of the substance. From this equation, the rate at which any body is radiating energy can be calculated if its temperature, T, and emissivity, e, are known. It is the value of the emissivity which varies with the material; σ always has the value given. For a perfectly black body, e has the value unity. Indeed, the emissivity of substances is defined in terms of the ratio at which they radiate energy compared to the rate of which a black body at the same temperature emits energy. That is, the emissivity of substances is given by

$$e = \frac{P^e}{P^e}$$

where P^e is the rate (in $\dfrac{\text{watts}}{\text{cm}^2}$ or $\dfrac{\text{joules}}{\text{cm}^2\text{sec}}$) at which substance x is

emitting energy and P^e is the rate at which a black body at the same temperature is emitting radiant energy. Thus, values of emissivity will always lie between zero and one. For highly polished aluminum at 25°C, for example, the value of e is 0.039. For powdered carbon, which approaches black body behavior, e can be greater than 0.9.

In addition to the emissivity, another important material property is the absorptivity, a. The absorptivity is defined as

$$a = \frac{P^a}{P^i} \tag{76}$$

where p^i is the incident radiation flux ($\dfrac{\text{watts}}{\text{cm}^2}$) and p^a is the amount of this incident flux which is absorbed. A true black body, which absorbs all incident radiation and hence reflects none, thus appears black and has an absorptivity of unity. Both (a) and (e) depend on the wavelength of the incident or emitted radiation, respectively, as well as on the angle and temperature at which this energy is being absorbed or emitted. It can be shown by pure thermodynamic reasoning, however, that the value of the absorptivity (a) for a particular wavelength, direction, temperature, and material is exactly equal to the value of (e) for that same wavelength, direction, temperatures, and material. A black body has (a) = (e) = 1 for all wave lengths, directions, and temperatures. Importantly, however, some materials can have a high absorptivity for incident energy of one wavelength and a low emissivity for energy being emitted at a different wavelength. A grey body is one which has a = e = (a constant less than one).

What wavelength of radiation will a substance at a temperature T emit radiation? Answer: At most wavelengths, but with the maximum energy (in w/m^2)/cm *not* (w/m^2)/cm^{-1}) being produced at the wavelength given by

$$\lambda_{\max} T = 2.892 \times 10^7 \ (\text{Å} \cdot {}^\circ\text{K}) \tag{77}$$

This equation is known as Wien's displacement law (Wilhelm Wien, 1864–1928, who won the Nobel prize in 1911 for this discovery). Thus, a solar energy collector which has reached a temperature of 400°K (127°C) will radiate its maximum energy at a wavelength of 72,300 Å. The sun radiates its maximum energy (in units of (w/m^2)/cm *not* (w/m^2)cm^{-1}) at about 5000 Å as may be calculated from Figure 72 by replotting this figure as (w/m^2)/cm versus cm. What is the temperature of the *light-emitting surface* of the sun assuming that it behaves as a black body?

As we shall see shortly, the fact that one material may have a high (a) for sunlight and a low (e) for its own radiation has tremendous consequences for solar energy collection. The invention of a *stable* material with a = 99% and e = 1% which would not degrade during heating would go a long way towards solving the solar energy collection problem.

In terms of collecting solar energy, we may now calculate the maximum temperature which a flat-plate, nonfocused collector can reach under the influence of sunlight. Suppose the intensity of this sunlight is $1000 \frac{\text{watts}}{\text{m}^2}$ and the area of the collector is one square meter. Assume that this collector absorbs 80% of the incident solar power, *i.e.*, it has an absorptivity of 80%. Assume further that the emissivity of this collector for long wave, infrared radiation is only 50%. Such a difference between absorptivity and emissivity is by no means impossible since the wavelength of the radiation being given off by the collector is far different from that of the sunlight it is absorbing, as we have just seen. Figure 75, for example, compares the solar spectrum with the spectrum emitted by a perfect black body (*i.e.*, e = 1) at 95°F.

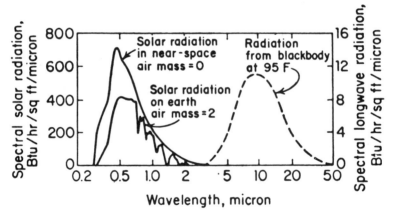

Figure 75. Comparison of the solar spectrum, (read units on the left ordinate) with that given off by a black body at 95°F (read units on the right ordinate). (Courtesy McGraw-Hill Book Co.)

Under these conditions, the amount of absorbed power will be 800 watts while the reradiated power from the absorber will be $\sigma e \, T^4$. The equilibrium temperature will be reached when the radiated power equals the absorbed power, neglecting convection losses and the thermal radiation of the collector environment. Thus,

$$800 \text{ watts} = \sigma e \ T^4 \tag{78}$$

or

$$T^4 = \frac{800}{(5.67 \times 10^{-8}) \ (.5)} \ ; \ T = 410°K = 137°C \tag{79}$$

Thus, the temperature of the absorber will only reach to a little above the boiling point of water. In even a perfect heat-engine operating between this temperature and an ambient temperature of 30°C, the efficiency would only be

$$\eta = \frac{T_H - T_L}{T_H} = \frac{413°K - 303°K}{413°K} = 26.6\% \tag{80}$$

Clearly, to increase this efficiency it would be necessary either to use focused collectors or to find a collector material with different values of a and e. As it happens, considerable effort has been devoted to focused collector design, but until recently, very little work has been done to develop materials with improved values of a and e.

If instead of a = 80% and e = 50%, the collector had a = 99% and e = 1%, the equilibrium temperature would rise to 1150°K (877°C)! That is the kind of temperature that is used in a modern-day steam plant. A material with these kinds of properties is called a selective absorber. Selective absorbers may be produced in the following way: As we have seen, polished aluminum has an emissivity of 0.039 for all wavelengths. A material such as CdTe (a semiconductor), has a high absorptivity for wavelengths less than about 8600 Å. For wavelengths somewhat longer than this, CdTe is as transparent as glass is to visible light. Thus, by placing this coating of CdTe onto aluminum foil one should be able to produce a composite material with a high absorptivity for sunlight but a low emissivity for long-wave infrared radiation. This is so because for this infrared radiation the CdTe is transparent; hence the reradiation properties of the material are, for this wavelength, determined primarily by the underlying aluminum foil.

Finally, to see the effect of focusing the sun's rays, imagine a parabolic collector like that shown in Figure 76. Assume that the concentration ratio is 10:1 and that a = 50% and e = 50% for the absorber, as before. The power density in 1000 w/m² sunlight is now raised to 10,000 w/m² on the absorber, and the equilibrium temperature of the absorber is increased to:

$$T^4 \text{ equilibrium} = \frac{8000}{(5.67 \times 10^{-8} \ (.5))} \ T = 456°C \tag{81}$$

It should be realized that as the degree of concentration increases,

Figure 76. A parabolic collector focusing sunlight onto a tubular collector at the Shuman-Boys solar power plant, 1913. (Annual Report, Smithsonian Institution, 1915.)

the importance of having a low concentration ratio may decrease relative to the importance of having a high absorptivity. This shift in relative importance will be true, for example, if heat is constantly being extracted from the absorber. Such a situation will be the case if the absorber is a pipe carrying water through the line focus of a parabola, like that shown in Figure 76, and this water is being turned into steam and used to drive a generator. If such is the case, the absorber temperature may be only 150°C even with a high concentration ratio, the excess power being drawn off as steam. The reradiation of energy from the absorber will then be relatively unimportant. The system shown in Figure 76 was operated during 1913-1914 in Meadi, Egypt and used to drive a 50-hp steam engine.

Traditionally, many focusing systems have utilized parabolic collectors. To obtain optimum results with a parabolic collector, however, it must be steered so as to be pointed at the sun. Such steering, involves substantial additional expense. Recently, however, a new collector design has been proposed by Dr. Roland Winston of the Fermi Institute in Chicago.

An example of one of these new collectors is shown in Figure 77. It consists of two segments, each of which is a section of a parabola. The axis of each of these parabolic segments, however, is not parallel to the axis of the resultant device. Such a collector has the property that it collects a higher proportion of incident sunlight over a much greater range of angles than does a parabolic collector. Hence, the need to steer such collectors has decreased. As with all collector systems, cost considerations are paramount. For special applications, where cost is no object, solar collectors have been used to produce temperatures in excess of 6000°C by means of concentration factors of more than 1000 to 1.

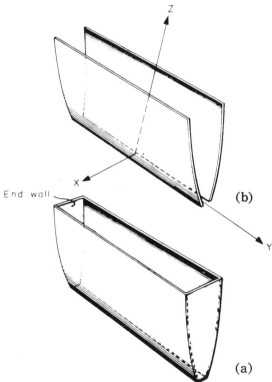

Figure 77. A modified concentrating collector, (a) consisting of two parabolic elements, (b) whose parabola axes are not quite parallel to the axis of the resulting collector. (Courtesy of Roland Winston and Pergamon Press.)

SOLAR HEATING AND SOLAR DISTILLATION

These are perhaps the simplest and most direct uses that can be made of solar energy. Indeed, in many parts of the world (Israel, Australia, Japan) solar water heaters and solar stills are already in use. The principle of solar water heating is extremely simple. Flat-plate collectors are used and incoming (cold) water is passed through this collector, before being pumped to an insulated storage tank. As shown in Figure 78, if the system is large enough this same hot water storage tank can also be used for supplemental home heating. Similarly, solar stills for the conversion of salt water to fresh water have long been in operation. In 1872 in Chile, for example, a 51,000-ft^2 distillation plant was constructed; this system is still in operation. In 1967, a 93,500-ft^2 facility was constructed to supply fresh water to the Greek island of Patmos. This plant can produce 7000 gallons of fresh water per day. Because of the simplicity of construction, construction costs are low. The 21,700-ft^2 facility on the Greek island of Messyros cost only \$2.11/ft^2 to build. The basic arrangement in all of these distillation plants is the same as shown in Figure 79. Sheets of glass are placed at an angle over shallow black basins of salt water. The sunlight passes through the glass without heating it and is absorbed by the water basin. This heat evaporates the water leaving behind the salt. The water vapor condenses on the cooler glass and runs into a collecting trough. The entire current engineering emphasis here is on lowering construction costs as far as possible.

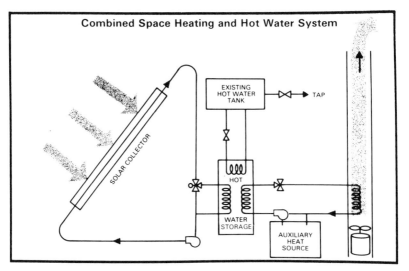

Figure 78. A schematic drawing showing the arrangement of a combined solar water-heater and supplemental solar space heating system.

Solar heating of homes has already been attempted in the United States, but only to a very, very small extent. One principal problem is the storage of heat. The overall arrangement of a typical system is shown in Figure 78. In principle, this heat storage may be achieved by using a sufficient volume of any material with a sufficiently high heat capacity. A typical home in the United States will require on average, approximately 1 x 10^8 cal/day for heating.

The specific heat of water is 1 cal/gram-°C or 1 Btu/lb°F. That is, it takes one calorie of heat to raise the temperature of one gram of

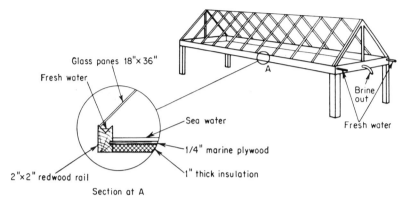

Figure 79. A schematic drawing showing the basic arrangement for solar powered water distillation. (Courtesy of McGraw-Hill Book Co.)

water one degree Centigrade (or Celsius); similarly, it takes one
British thermal unit (Btu) of heat to raise the temperature of one
pound of water one degree Fahrenheit. The density of water is 1
gram/cm^3 (62.4 lb/ft^3). One single cubic foot of water can thus store
624 Btu or 1.57 x 10^5 cal if its temperature is raised 10°F (5.56°C). A
3000-gallon storage tank and a temperature rise of 25°F would provide 625,625 Btu or 1.58 x 10^8 calories or somewhat more than that required for one day's heating. This amount of water would weigh
about 12 tons, and occupy a volume of 401 ft^3. If more than
a one day reserve were required, the mass of water would go up
proportionately. Massive amounts of crushed rock (with a specific
heat of 0.2 Btu/lb°F) could also be used. Clearly, some better means
of storing heat would be extremely useful. Rather than relying on
specific heat, one might utilize the heat of fusion of low melting point
materials instead. The heat absorbed when $CaCl_2 \cdot 6H_2O$ melts (at
about 34°C) is 75 Btu/lb. Thus, one pound of $CaCl_2 \cdot 6H_2O$ can store as
much heat as one pound of water heated to 75°F. Actually, $CaCl_2 \cdot$
$6H_2O$ can store more than this if it, too, is heated above its melting
point. $CaCl_2 \cdot 6H_2O$ is, however, much more expensive than H_2O or
rocks. What is needed more than anything else to make solar heating practical is a cheap way of storing large amounts of low-grade
(*i.e.*, low-temperature) heat. Alternatively, better collectors which
could give higher temperatures, would allow the storage of heat as
steam. When steam condenses, it gives off a huge amount of heat
(540 cal/gram).

SOLAR COOLING

One great advantage of using solar energy for air-conditioning
and refrigeration is the fact that the need for such refrigeration will
generally be highest when the supply of solar energy is also highest.

Hence the need for the large scale storage of thermal energy is intrinsically not as severe for solar refrigeration systems as for solar heating systems.

How exactly can solar heat be used to power a refrigerator? The principle is exactly the same as that which has been used for many years in gas refrigerators. As an example, consider ·refrigeration using ammonia. Liquid ammonia (NH_3) is evaporated from a vessel and, after passing through a tube, is dissolved into water to form a low-pressure solution of ammonia (NH_3OH). Remembering that it took 540 cal/gram to turn water to steam at its boiling point, it should now be noted that it takes approximately 300 cal/gram to turn (NH_3) liquid to (NH_3) gas. Therefore, in evaporating and changing from liquid to gaseous ammonia, the chamber originally containing the liquid ammonia is cooled. The recharging of this refrigeration system with solar energy proceeds as follows: focused solar radiation is used to heat the aqueous ammonia (NH_4OH) solution. As the solution is heated the solubility of the ammonia decreases and this high pressure ammonia condenses in the starting chamber. This chamber is, of course, heated up as the 300 cal/gram heat of condensation are given off. This heat can be removed however, by radiation and conduction. The liquid ammonia at room temperature and under pressure, can then be used for cooling again. For continuous operation, two circulation loops are needed. One loop consists only of ammonia, whether gas or liquid. The second loop consists of ammonia in solution. Also present in real systems is hydrogen gas, which is practically insoluble in water, does not condense to liquid in the system, and is used to achieve an overall pressure balance and prevent the water and liquid ammonia from mixing. This continuous process is called the Electrolux absorption refrigeration system and was invented in the 1920s by two students, Carl Munters and Baltzar von Platen, while they were undergraduates at the Royal Institute of Technology in Stockholm.

If such refrigeration systems work then why are they not widely used to achieve cooling via solar energy? The answer is simply that focused solar energy collectors such as that shown in Figure 75, are too expensive. Also, of course, some storage capacity must also be provided for cloudy, but hot, days. Currently available flat-plate collectors do not produce a high enough temperature to drive the ammonia out of its pressurized water solution. As you may already see, there are two different possible solutions to this impasse. One solution is to find some other system besides $NH_3 - H_2O$ for the refrigeration system. A good deal of work has been done, for example on the system sodium thiocyanate-ammonia as well as on the system lithium bromide-H_2O. If a system could be found that would work effectively with temperatures of less than 100°C, then air conditioning systems could be built using currently available flat-plate collection systems. A second solution would be the development of better selective absorber coatings. As we have already seen, a selective absorber surface with an a/e ratio of 99 could reach over 800°C in full sunlight. At such temperatures, normal H_2O-NH_3 absorption refrigeration systems could work very effectively.

The problem of energy storage in solar powered air conditioning systems is not, as we have seen, as demanding as the problem of energy storage in solar heating. In addition, two quite distinctly different approaches to the storage problem are possible. One may seek to store either the collected solar energy of as high a temperature as possible or one may store "coolness," *e.g.*, by producing ice which absorbs 79 cal/gram during melting. In either case, the best storage means will almost certainly involve materials which undergo a phase change, with either accompanying absorption or release of heat. The whole question of phase changes in materials is somewhat beyond the scope of us here, and is best dealt with in a course in the science of materials. It is worth noting, however, that only a very tiny fraction of possible heat or cold storage materials have been investigated. It may even be possible to custom design new materials which have exactly the right properties. This is a virtually unexplored field.

LARGE-SCALE ELECTRIC POWER GENERATION

Solar generation of electricity for use on earth holds the promise of an abundant, clean, inexhaustible source of power. There have, of course, been numerous approaches to the problem of converting solar energy into electricity.

The use of solar radiation for central power generation will ultimately depend upon economics. Solar power plants will be capital-intensive. Since the fuel for the plants is free, the cost of the resultant power will depend directly on the cost of the system required to convert sunlight to electricity. Typical current producer costs of electricity are about $0.01/kwh. If storage methods to enable solar plants to supply electricity on a 24-hour basis cannot be produced, the value of the electricity produced by solar energy will be worth less than half of that from a 24-hour plant. This result occurs because interruptible power must compete with lower cost "dump" power from conventional plants which are not being fully utilized. Currently, electricity produced from solar energy costs approximately 100 times more than that from coal fired central power stations. Such a cost cannot be borne except in very special situations. In this section we will consider two principal methods for achieving the conversion of solar energy to electric energy. These two methods are solar thermal conversion and photovoltaic solar energy conversion. Other, natural, means of solar energy collection occur in the atmosphere, in the oceans, and on land, giving rise to wind, rain, ocean currents, ocean temperature differences, and the growth of plants. The use of the first of these resultant effects in producing electricity are discussed in Chapter 10. In this section we will consider two fundamentally different methods of large scale electric power generation from sunlight. These methods are, first, the direct conversion of sunlight into direct current electric power via photoelectric effects and, second, the use of thermal systems to produce electric power via relatively straightforward thermodynamic processes.

In photovoltaic energy conversion, the basic problem concerns the nature of the device itself. Solar cells (or solar batteries, which is strictly speaking, incorrect) are semiconductor devices which work as follows: The incident sunlight breaks covalent bonds in a semiconductor. In a solar cell a p-type semiconductor (one with not enough bonds) and an n-type semiconductor (one with too many bonds) are in direct contact. If light with a short enough wavelength (high enough energy) shines on the junction between the p and n-type semiconductors, it will break some bonds in both the n and the p-type semiconductors. One broken bond produces one free (unbonded) electron and one hole (lack of an electron bond). Of particular importance is the local field gradient which exists near the junction of the p and n-type regions. The interdiffusion of holes and electrons in this region creates a local field which causes the separation of the holes and the electrons generated by the incident photons before they recombine and eliminate each other. The separation of these charges produces a voltage, measurable in an external circuit. This voltage can be used to produce a current in the external circuit. Figure 80 gives a cross-sectional view of a typical silicon solar cell. Such cells are currently used routinely to supply electric power for spacecraft. A proposal by Peter Glaser of Arthur D. Little, Inc., is that a large array of such cells in space be used to produce, continuously, electric power which could then be transmitted back to earth using microwaves. Alternatively, such cells could be used directly on earth, where the sun, however, does not shine continuously. In either case, however, the key element is the cost of the cells.

Currently, solar cells based on silicon can be produced which can convert sunlight to electric power with up to 18% efficiency. The cost of these cells is, at best, about \$20,000/kw. Fossil-fueled electric power plants can be built for about \$300–\$400/kw. Thus, the key problem here is how to make solar cells cost about 1/100th of what they do now. There are two approaches to this problem. One approach is to try to produce silicon cells more cheaply by growing the silicon crystals directly in thin sheet form. The other approach is to try to make thin film cells, not of silicon. $CdS/Cu_{1-x}S$ cells can, for example, be made by vacuum deposition of CdS onto aluminum foil followed

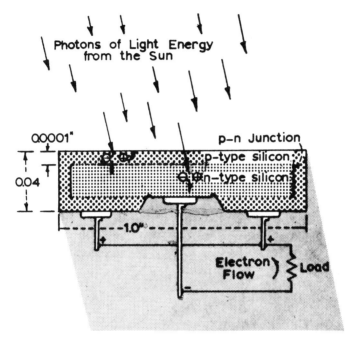

Figure 80. Cross-sectional view of a single crystal silicon wafer prepared as a solar cell. (Courtesy of Bell Laboratories Record.)

by chemical preparation of a surface $Cu_{1-x}S$ layer in cuprous chloride solution. The overwhelming advantage of thin-film solar cells is that they can be made using relatively little material. This is important if one is talking about building enough solar cells to cover square miles of area. Currently, the best thin-film cells are 5% efficient but, unfortunately, are not stable in the presence of sunlight, moisture, and air. A Nobel Prize awaits the person who can make cheap, stable, efficient solar cells.

Another possibility that should be mentioned is that of the combined production of a selective absorber surface and a thin-film solar cell in a single unit. As we have seen previously, the best selective absorber surfaces consist of a thin-film semi-conductor layer on top of a bright metal surface. Such an assembly is halfway towards becoming a solar cell. It would seem possible to make a thin-film photovoltaic cell which was also a selective absorber surface. Remember, though, that as the temperature rises, the efficiency of a solar cell fails. Thus, the decision as to what temperature at which it is best to operate the combined solar cell-selective absorber, becomes a problem in optimization theory.

To overcome all problems associated with clouds or the day-night cycle, it has been proposed that photovoltaic devices be arrayed in space and the resultant electric power transmitted to earth in microwaves. Satellites can be placed in synchronous orbits such that they remain over the same point on the earth but still receive solar energy 24 hours a day. The use of microwaves having a wavelength of 1 cm for power transmission would keep atmospheric losses to about 6% and have them remain moderate even in severe rainstorms. A schematic drawing of this system is shown in Figure 81. To transmit 10,000 Mw would require a 1-km (0.62 miles) diameter transmitting antenna and a 7-km (4.35 miles) diameter antenna, to keep the transmission efficiency (DC to DC) to between 55 and 75%. The installation of such a system would require an operational space shuttle system. The projected cost for such a system is about the same as that of the entire Apollo program. The building of such a system would require a major national commitment. It would be possible, of course, to build a much smaller demonstration system first without such a national commitment.

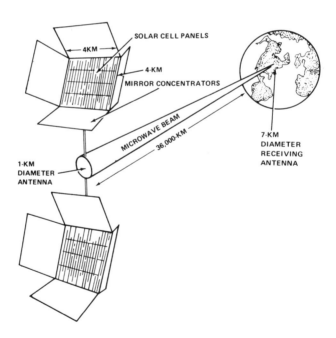

Figure 81. A schematic drawing showing the approximate solar cell panel and antenna dimensions required, with present technology, to produce 10,000 Mw of electrical power on earth using solar cells in synchronous earth orbit.

Thermal methods for the conversion of solar energy to electric energy have several factors in common with those considered in the care of solar heating or cooling. That is, the solar energy must first be collected as heat, with emphasis as before on achieving temperatures as high as possible. The use of selective absorber surfaces may allow the achievement at temperatures in the range of standard steam turbogenerators (typically 565–595°C) with solar concentration ratios as low as 10 to 1. One pressing problem has been that currently available selective absorber materials are not stable at these high temperatures. The absorptivity slowly decreases while the emissivity slowly increases during use and the resultant maximum temperature consequently decreases. The most favored designs currently envision "linear heat-absorbing systems," *e.g.*, pipes running along the length of long collectors such as that shown in Figure 76. A large-scale solar thermal power plant could be built today. The principal question is whether such a plant would be able to compete commercially with other methods of power generation. Without the use of selective absorbers, the answer to that question is no. To reach high concentration ratios required for high temperature without selective absorbers, the cost of the absorption collection system and its associated required steering system, become prohibitive. The key limiting factor then becomes the stability of the selective absorber material.

As we saw earlier, if the working temperature of the absorber is kept relatively low, the need to maintain a low emissivity is decreased. A low working temperature means a low thermodynamic efficiency and hence a larger collector area to produce one kilowatt of electrical power. The problem thus becomes one of optimization-absorber lifetime versus collector area. While the needed collector area can be easily calculated, there is only very scattered and incomplete data available on the stability of selective absorbers. One obvious problem is the difficulty of rapidly determining long-term material stability.

OTHER METHODS FOR SOLAR ENERGY CONVERSION

Plants may be grown using sunlight and then burned to produce electricity by means of normal thermodynamic cycles. Up to 26 tons of dry plant material may be produced per acre per year. Such plant material may produce up to 16 x 16[6] Btu/ton. The overall solar energy conversion efficiency, to stored Btus, may be 3.0%, which is not very high. Also, there is a large manpower requirement in harvesting this plant material. Even so, however, 3% of the land area of the U.S. land area could produce the stored Btu energy equivalent of all the electric energy anticipated for 1985. One related approach is to combine algae or other plant growth with sewage processing. Algae may be grown in shallow ponds with added animal wastes. In this way, more combustible matter may be produced per acre than by normal earth farming. Water hyacinths, for example, will produce up to 85 tons per acre of dry product.

The only area for novel work appears to be the development of marine areas for salt-water plant culture. Only in this way does it appear possible to decrease the "land" cost of the plant. Also, it might be possible to breed more productive sea plants than those which now exist.

In terms of tree farming (and the so-called solar plantation), it has been calculated that for steady-state operation, a 1000-Mw plant, a typical current plant size, would need a supply area of about 500 square miles of trees. The estimated fuel cost from the resultant timber is $2/10^6$ Btu. The main problems with this concept are the need for cheap, fertile land and the need for low-cost methods of harvesting, processing, and transporting the material.

The H-O bond in water can be broken directly by ultraviolet light with a wavelength of 2500 Å or less. Since the H and O nuclei (connected by this bond) are vibrating, however, some bonds may be broken by longer wavelength light (lower-energy light) than this. In principle, most of the ultraviolet wavelengths of sunlight could decompose water into hydrogen and oxygen. Water is not normally decomposed by light, however, because the bond immediately reforms. Some materials *e.g.*, TiO_2, appear to act as catalysts and enable two (H_2O) molecules to be decomposed to $2H_2$ and O_2. This hydrogen and oxygen can then be recombined in a fuel cell to directly produce electrical power, or burned in more usual engines. There are several groups now working in the U.S. to try to find a better catalytic material than TiO_2, but little is really known of the current status of this materials science research. This method of solar energy utilization is extremely interesting because of its extreme (in principle) simplicity. The catalyst would be placed in the water and the water then exposed to sunlight and the (mixed) hydrogen and oxygen collected. Clearly, energy storage is no problem in such a case. Unfortunately, current efficiencies of conversion of solar energy to chemical energy ($H_2 + O_2$) are extremely low (much less than 0.1%). It is difficult, if not impossible, to predict whether more efficient conversion materials will be discovered. This method has such possibilities; however, it deserves more work.

Lastly, consider the use of the rather unusual alloy, Nitinol, in solar energy conversion. Nitinol is short for nickel-titanium-Naval Ordinance Laboratory. Nitinol, and a few other alloys have the very unusual property that they "remember" their previous shape. Consider a Nitinol wire that is initially straight at a low temperature, T_L. Let this wire be heated to above a certain very high temperature, and at this high temperature, 1000°C for example, let it be bent into a U-form. This is the shape it will remember. Then let it be cooled down to a temperature less than the transition temperature (T_t). Suppose, for example, the wire is cooled back to T_L. At T_L, let this wire be bent back to a straight wire. If this restraightened wire is then heated to T_t or higher, it will spontaneously reform the U shape.

The explanation of this behavior is a little involved and is best given completely in a course in physical metallurgy and material science. Basically, however, the metal is in one crystallographic form at T_t and in another form at T_L, and the transition path from one form to the other is related to the microstructure developed during the original high temperature bending operation. How then may this effect be used in solar energy conversion?

When the right alloy composition is chosen, a typical value for T_t can be 85°C, and a typical value for T_L can be 25°C. Thus, such an engine could run using the temperatures produced by currently available flat-plate collectors. Secondly, and most importantly, the work required to straighten the wire at T_L is less than the work that can be produced when the wire rebends itself above the temperature above the temperature T_t. Such wires may thus be assembled into a heat-engine which operates between relatively closely spaced temperatures. The maximum efficiency of such an engine will therefore be relatively low:

$$\eta_{max} = \frac{(85 + 273) - (25 + 273)}{85 + 273} = 16.7\%$$

Actual efficiencies will, of course, be considerably lower than this. Even so, however, the simplicity of the resulting device has many attractions. This idea is so recent that there has not, as yet, been any major attempt made to apply it to practical systems. There may be practical difficulties which are not yet apparent.

The attractions of solar energy are so great and the possible rewards so large that there will surely be an increasing number of imaginative and innovative attempts made to harness this inexhaustible power source. In this chapter we have only been able to outline those methods being currently pursued. There will surely be many new developments in future.

SUGGESTED FURTHER READING

1. Wolf, M. "Solar Energy Utilization by Physical Methods," *Science* **184**, No. 4134, 382 (1974).
2. *Proceedings of the World Symposium on Applied Solar Energy,* Phoenix, Arizona, November 1–5, 1955, (Stanford, California: Stanford Research Institute, 1956).
3. Daniels, Farrington. *Direct Use of the Sun's Energy* (New Haven, Connecticut: Yale University Press, 1967).

194   *Introduction to Energy Technology*

4. Hottel, H. C. and J. B. Howard. *"New Energy Technology Some Facts and Assessments,"* (Cambridge, Massachusetts: The MIT Press, 1971).
5. Zarem, A. M. and D. D. Erway. *Introduction to the Utilization of Solar Energy* (New York: McGraw-Hill Book Co., 1963).
6. Yellott, J. J. "Solar Energy," In *Standard Handbook for Mechanical Engineers*, T. Baumeister, Ed., 7th ed. (New York, N. Y.: McGraw-Hill Book Co.).
7. Rau, H. *Solar Energy* (New York, N. Y.: Macmillan, Inc., 1964).
8. *Solar Energy as a National Energy Resource*, NSF/NASA Solar Energy Panel Report, Document PB 221659 (Springfield, Virginia: National Technical Information Service, 1973).
9. Federal Energy Administration. "Project Independence Report on Solar Energy" (Washington, DC: U.S. Government Printing Office, 1974).
10. Williams, J. Richard. *Solar Energy Technology and Applications* (Ann Arbor, Michigan: Ann Arbor Science Publishers, Inc., 1974).

PROBLEMS

1. The sun is 93,000,000 miles from the earth and is 880,000 miles in diameter. It delivers 1.36 kw/m² onto the earth. Assume that the emissivity of the sun is unity. Calculate the effective surface temperature of the sun.

2. Explain the fact, as shown in Figure 73, that the seasonal variation in solar power for La Jolla, California, is substantially less than for Fresno, California.

3. The electric power requirement of the New England states is approximately 15,000 Mw, and this power level represents the average level over a 24-hour period. If sunlight can be converted to electrical power with only 10% efficiency, what fraction of the area of the state of Arizona would have to be covered with solar cells to supply the entire New England electrical energy requirement. Of course, large-scale storage schemes would have to be devised to store energy during the 12-hour period when solar energy is not being received in Arizona. Also, transmission of this power would be a major, but not impossible problem. Ignore these difficulties; assume it can be done. Hint: Arizona receives an average of 2000 Btu/ft² of solar energy per day.

4. A solar heating system is being designed for a house. To store energy during periods of cloudy weather, a salt mixture which melts at 60°C is to be used. In the region where the house is located, the sunlight falling on the roof of the house delivers at least 500 w/m², even in wintertime. A flat, selective absorber system is to be used to collect the solar energy and melt the

salt. If the absorptivity of the selective absorber for sunlight is 95%, what must be the emissivity for reradiated power in order to melt the salt? Neglect any heat losses other than reradiation.

5. Calculate the temperature, in °C, that a perfect black body would reach at noonday in Arizona on a day when there is no wind and the intensity of sunlight is 1100 w/m².

6. Calculate the temperature which could be reached with a selective surface, flat-plate collector whose absorptivity for sunlight was 99%, and whose emissivity for reradiated power was 1%.

7. Current commercial silicon solar cells cost as little as 25¢ per square centimeter and can convert sunlight into electricity with at least 10% efficiency. It currently costs about $150,000,000 to build a 750-Mw, coal-fired electric station. What would be the cost of the solar cells needed for a 750-Mw plant in a region where the intensity of sunlight is 900 w/m²?

8. Assume silicon solar cells cost $0.25/cm². If the cost of a collector to concentrate sunlight onto these cells cost $1.25 $^x/_{10}$ where x is the concentration ratio of the collector, what would be the cheapest system to produce 750 Mw in sunlight whose intensity is 900 w/m²? The cells are 10% efficient and this efficiency is not affected by sunlight concentration.

9. Discuss the probable impact on the rest of the world if the United States could develop photovoltaic cells which provided electricity from sunlight more cheaply than electricity can be produced from coal.

10. A solar concentrator is used to send focused sunlight having an intensity of 10,000 w/m² onto a collector having an area of two square meters. This collector has an absorptivity of 90% for sunlight and an emissivity of 10% of reradiated power. From this collector, 1500 Btu/hr are being taken away by circulating water that is being used to power a steam engine. At equilibrium what is the temperature of the collector?

11. What do you think would be the most practical steps the government could take to develop the use of solar energy on a large scale? Justify your answers.

Chapter 10

Wind, Tidal, Geothermal, and Ocean Thermal Gradient Energy

INTRODUCTION

In this chapter we will consider four special sources of energy. Three of these—winds, tides, and ocean thermal gradients—are renewable and can be expected to last into the foreseeable future. The fourth, geothermal energy, might become exhausted in local regions under particular conditions, but in its entirety could not possibly be used to exhaustion. In each case, very special engineering, economic, and preexisting, natural considerations may determine whether exploitation is possible. In several instances no new technical developments would appear to be needed. Only imaginative engineering and political leadership appear to be required to bring about commercial electric power production. In each of the following sections, we consider first the general nature of the energy resource before going on to discuss specific engineering and economic problems.

WIND ENERGY

Wind power cannot be expected to solve all our energy problems. As we will see, however, with enthusiastic, large-scale exploitation, wind power systems might provide a significant fraction of our energy needs. Even with conservative estimates the harnessed energies of the wind could provide substantial energy in favorably situated locations at costs which do not appear to be prohibitive. Wind power has, of course, the great advantage that it is a self-renewing energy resource which would provide a continuing supply of nonpolluting energy for the foreseeable future. In the chapter on solar energy, we saw that about 30% of the incident solar energy never reaches the surface of the earth, but is absorbed by the atmosphere. The resulting circulation of the earth's atmosphere can give wind velocities of

197

up to 200 mph blowing at altitudes between 20,000 and 40,000 ft above sea level. Planes traveling from the east coast to the west coast of the U.S. use these winds to increase groundspeed. Thus, transcontinental airline times are usually shorter east-to-west. In terms of electrical power generation, however, what is of primary importance is the energy of air streams within a few hundred feet of the ground. Even so, however, there are many locations where the wind direction and velocity are relatively constant over long periods, as may be seen from Figure 82.

Figure 82. Photograph taken at Kitty Hawk, North Carolina, at the site of the Wright brothers' first flight, illustrating the constancy of the prevailing winds at this location.

For centuries, the energy of the wind has been harnessed in a variety of simple ways. The possible harnessing of wind for modern-day uses has now stimulated a variety of new schemes, and we shall examine some of the more promising of these.

Windmills have been used to convert wind energy into mechanical energy for more than 700 years, and since 1890 windmills have been used together with generators to produce electricity for day-to-day needs. Periodically, larger-scale wind-to-electrical power conversion projects have been attempted. A 1931 Russian system at Balaclava, for example, generated for several years about 250,000 kwh/yr for the Yalta electric grid. The rotor of this machine was 100 ft in diameter and it was on a tower 100 ft high. Currently, the most successful system has been operated near Gedser, Denmark. This fully automated 200-kw wind turbine has three blades each 40 ft long mounted on a 75-foot tower, and has produced about 400,000 kwh/yr. During the years 1941–1945, a 1250-kw wind-powered generator on Grandpa's Knob, near Rutland, Vermont, produced power for the Central Vermont Public Corporation. This wind-powered turbine had a two-bladed propeller 175 ft in diameter, as shown in Figure 83. The pitch of these blades could be regulated during operation to control the power to the generator. They could also be feathered (turned edgewise to the wind) to avoid damage during gales. This system was used to produce power in 70-mph winds and withstood winds of up to 114 mph. It failed on the morning of March 26, 1945, when the spar for one of the blades suddenly broke at the root and the entire 8-ton blade was hurled 750 ft. The vibration caused by the remaining blade would have destroyed the tower had not the foreman reacted quickly and feathered the remaining blade. Subsequent failure analysis showed that stress concentrations due to improper welds at the blade roots had led to progressive (presumably fatigue) cracking. The failure of this system was unfortunate because it effectively ended interest in wind power for 30 years. It was the work of essentially one man, an engineer named Palmer Putnam, who single-handedly brought together the required people, money, and engineering plans. The mechanical failure of this system and the subsequent demise of the entire project must have been a great disappointment to him.

Interest in wind energy systems has recently been revived by William E. Heronemus, a Professor of Engineering at the University of Massachusetts, Amherst. Rather than using mountain tops, Professor Heronemus has proposed building wind turbines on the Great Plains area of the midwest. According to his calculations, a system of 300,000 wind turbines could produce the equivalent of 189,000 Mw of installed nuclear power plant capacity. The wind-powered system would have a much larger total capacity. The 189,000 Mw would be the calculated minimum power output. The total 1970 U.S. installed capacity was 360,000 Mw, so the Heronemus plan would certainly be of considerable size.

As another example, it has been estimated that the total average wind power over Wisconsin averages 3.53×10^{12} w, which is far more power than is now used in Wisconsin for all human purposes. Wisconsin has a surface area, however, of 5.47×10^{11} square miles or $1.42 \times 10^{11} m^2$ so that the power density is only about 25 w/m^2 which is about 40 times less than the maximum solar power density. As you would expect, however, certain regions will have a much higher

Figure 83. The Putnam wind turbine, America's largest windmill, which generated power for a Vermont utility company in the early 1940s. (Courtesy of Carl J. Wilcox Associates.)

average wind power density than this average because of local climate and terrain conditions. As an example, Figure 84 shows how smooth, rounded hills can act to increase the wind power density, while rough surface terrains may produce unusable turbulence. The smooth hill, by increasing the velocity of the laminar, *i.e.*, evenly flowing, wind can greatly increase the usable wind power density.

The NASA-Lewis Research Center in Cleveland, Ohio, is currently managing several large scale wind experiment projects aimed at the development of reliable and cost competitive wind energy conversion systems. The first device to be built is a 100-kw to be installed at Plum Brook, near Sandusky, Ohio. This device will have a rotor 125 ft in diameter and will begin turning in 8-mph wind, and be capable of generation in winds up to 60 mph.

To estimate the power in wind, remember that power is energy per unit time. The kinetic energy of any particle is, of course, ½ MV². The volume per unit time of wind, moving with a velocity V and pass-

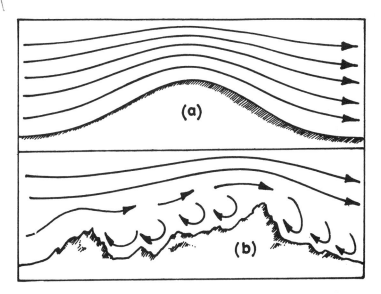

Figure 84. The effect of surface topography on surface wind patterns: (a) over smooth hills, wind speeds increase, (b) over rough terrain, turbulence develops.

ing through a given area A is AV. The mass per unit time is ρAV where ρ is the air density (kg/m³). The power (kinetic energy/unit time) of wind, having a velocity V and a density ρ passing through a propeller of area A is, therefore,

$$\text{Power} = \frac{\text{½ MV}^2}{\underset{\text{time}}{\text{unit}}} = (\tfrac{1}{2})\,(\rho AV)\,(V^2) = \tfrac{1}{2}\,\rho AV^3 \qquad (82)$$

Thus, the power in wind increases with the *cube* of the wind velocity. Current propeller and generator designs can convert only a small percentage of this power into electricity. One area, therefore, which deserves more study is that of improved propeller-generator systems. It would appear reasonable to expect to be able to improve this figure substantially. It is not possible, however, to utilize all the power contained in wind. Professor Betz at Gottingen in 1927 was able to show that the maximum power (not energy) that any windmill could ex-

tract from a wind stream was only 16/27 or 59.3% of the power contained in this wind. He reasoned in the following way:
Let

$$V_1 = \text{the wind speed at a long distance upwind}$$

$$V = \text{the wind speed pulling through the rotor}$$

$$V_2 = \text{the wind speed at a long distance downwind}$$

If N(kg/sec) of air pass through the rotor then the rate of change of momentum is $N(V_1 - V_2)$, which is equal to the force (thrust) exerted by the wind on the rotor. The power extracted from the wind by the rotor is therefore $V(N(V_1 - V_2))$ $\dfrac{m}{sec}\ \dfrac{kg}{sec}\ \dfrac{m}{sec}$ or $NV(V_1 - V_2)$ watts. Alternatively, the rate of change of wind kinetic energy is $\frac{1}{2} NV_1^2 - \frac{1}{2} NV_2^2$. Equating these two expressions gives

$$N(V_1 - V_2)\,V = \tfrac{1}{2}N(V_1^2 - V_2^2) \qquad (83)$$

or

$$V = \frac{V_1 + V_2}{2} \qquad (84)$$

since $N = \rho AV$, the power extracted by the rotor may be rewritten from $M(V_1 - V_2)V$ to $\rho AV(V_1 - V_2)$ where $\rho = $ the air density (kg/m³) and $A = $ the area swept by the rotor (m²), with the value of V just given, this becomes

$$P = \rho A \frac{V_1 + V_2^2}{2}\,(V_1 - V_2) \qquad (85)$$

From this expression it can readily be shown that the maximum power absorbed by the rotor occurs when $V_2/V_1 = 1/3$ and that this maximum power is 16/27ths of the power in the incident wind. Subsequent analyses which include the effects of downstream turbulence have raised this figure slightly, but no windmill can be more than about 60% efficient. This efficiency is still far higher than, for example, a typical windmill's 5% conversion efficiency. Even with a 5% figure, however, current estimates of the economics of wind power in favorable wind locations put this cost at about $325 per installed kilowatt in comparison with about $300/kw for nuclear reactors and $240/kw for coal-fired plants. Clearly wind power does not appear that impractical. What appears to be needed most is interest, determination, money, and good engineering designs.

Everyone has probably had in his mind's eye a picture of a wind generator system involving some sort of wind turbine with a horizontal axis as shown in Figure 83. This arrangement has, of course, been the traditional arrangement for windmills. There appear to be some definite advantages, however, for considering wind power devices which have a vertical axis. One such vertical axis device is the Savonius S-rotor. This device consists of a vertical cylinder sliced in half from top to bottom, the halves then being separated by about 20% of the diameter, as shown in Figure 85. This design can possess a high efficiency (31%) but is inefficient per unit of weight. This is so because all the area swept by the wind is occupied by metal. For a machine to produce 1000 kw in a 30-mph wind a Savonius machine requires about 30 times as much metal as a two-bladed turbine. The principal advantage of the S-rotor is that its action does not depend upon the horizontal direction of the wind. This feature greatly simplifies the design of the rotor support mechanism.

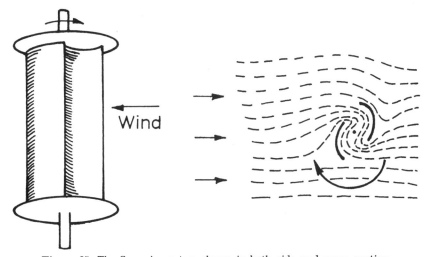

Figure 85. The Savonius rotor, shown in both side and cross section.

The Darrieux rotor is an alternate type of vertical axis machine. As shown in Figure 86 the Darrieux rotor obviously requires considerably less metal than the Savonius rotor, and yet is, of course, nondirectional. Its disadvantages are that rotation will not begin in wind velocities of less than about 12 mph. The aerodynamics of the Darrieux rotor are not simple. For example, it has been reported that no matter what the wind velocity, a simple Darrieux rotor will only reach 13 rpm. If it is directly accelerated to 65 rpm, however, the rotor will itself accelerate to over 200 rpm. Current 15-foot designs will produce 1.3 hp in a 15-mph wind. Because of their complex behavior, it may be confidently assumed that with a concentrated engineering effort, improved Darrieux designs should be possible. The trial testing of such designs using the wind tunnel in the engineering building addition, would be an excellent student project. In current Darrieux designs, for example, the vanes are of uniform cross section. This is almost certainly not the ideal shape. The determination of the ideal vane configuration is an important fluid mechanics problem that is well worth solving, using either theoretical or experimental methods. In addition, simple Darrieux rotors are not self-starting and designs to enable the device to begin rotation by itself have recently been announced. These designs essentially consist of the addition of starter buckets similar to the Savonius S-rotor.

One result of the application of fluid mechanics techniques to wind power, is the conclusion that substantial performance advantages can be obtained by using a shroud and diffuser on a wind turbine. The arrangement is shown in Figure 87. The basic function of this arrangement is to enable the rotor to capture airflow from a free

Figure 86. A Darrieux rotor, built by NASA's Langley Research Center, Hampton, Virginia. (Courtesy of the American Society of Mechanical Engineers.)

stream area that is greater than that of the rotor itself. Of course, a ducted rotor is never better than a free rotor that has an area equal to the area of the duct exit. Nevertheless, a ducted rotor can have a lower wind velocity cut-in speed than a nonducted rotor, as well as capturing more power than a nonducted rotor of equivalent size.

Figure 87. An artist's drawing of a diffuser-augmented wind turbine.

With regard to the design of the blades of wind turbines, one economical possibility is the use of flexible sailwings, as shown in Figure 88. Such wings have a greater ratio of lift to drag for shallow angles of attack than rigid wings, because of the flexibility of the sailwing surface. The use of sailwings is limited, however, to rotor diameters of 50 ft or less, because of the development of instabilities with high rotor-tip velocities. These same instabilities also limit the windspeeds in which such sailwing wind turbines may be used.

At a last example of alternate approaches to windmill design, consider the Magnus effect. A curve ball can be thrown only because of the Magnus effect, which is also manifested in the sideways thrust exerted by a cylinder spinning in a wind stream.

This effect was utilized by a German inventor, Flettner, who constructed a ship which successfully crossed the Atlantic in 1925 using the Magnus effect. A year later Flettner built a windmill, the four blades of which were tapered rotating cylinders with a diameter of 65 ft 8 in, driven by electric motors. This windmill produced 30 kw in a wind of 23 mph. A complete analysis of the fluid dynamics of a Magnus effect windmill has never been made. It is not known, for example, what would be the optimum cross sectional shape of the rotating cylinder-like members. The maximum efficiency of such Flettner windmills is also not known.

There are clearly many unanswered questions in wind turbine technology. It is possible that the development of new designs based on old concepts or of substantially new techniques could radically improve the possibility of utilizing windpower on a large scale.

Figure 88. A sailwing windmill in use in Wisconsin. (Courtesy of Ted Rozumalski/ Black Star Publishing Co.)

TIDAL ENERGY

The tides offer a virtually inexhaustible natural source of energy which is essentially unused, although the idea of harnessing the rise and fall of the oceans has long attracted engineers and inventors. The simplist tidal power device is a tide mill, a variety of water wheel utilizing the inward and outward flow of tidal water. The use of such devices goes back into antiquity. *Doomsday Book* (11th cent.) mentions what may have been a tide mill at the entrance to the Port of Dover, England. Other tide mills have been built from time-to-time along the coast of Great Britain and along the west coast of Europe. The first tidal mill in the United States was constructed in Salem, Massachusetts, in 1635. Another tidal mill in Rhode Island had 20-ton wheels 11 ft in diameter and 26 ft wide.

The origin of all tidal energy is the kinetic and potential energy of the sun-earth-moon system, but principally the kinetic and potential energy of the rotating earth and the orbiting moon. The level of the oceans rises and falls in response to changing gravitational forces. Therefore, this energy source is not, strictly speaking, inexhaustible. However, for all practical purposes, it can be treated as self-renewing. In any event, the harnessing of tidal power by man will not deplete this energy source any sooner than it will be depleted naturally. If not harnessed, the energy of the tides is lost due to friction between the moving water and the ocean floors. Interestingly, approximately half of all tidal energy dissipation occurs in the shallow Bering Sea. One effect of these friction losses is to slow down the rate of rotation of the earth. Over the past 120,000 years, the length of each day has increased by approximately one second. From this information the total amount of tidal power can be very crudely estimated.

The kinetic energy of a solid sphere of radius, R (m) with a mass M (kg), rotating with an angular velocity W (rad/sec), is given by

$$\text{Kinetic Energy (Joules)} = \frac{1}{5} MR^2 W^2 \qquad (86)$$

With this relationship plus the fact that the mass of the earth is 5.98 x 10^{24} kg and that tidal action has slowed the rate of rotation by one second in 120,000 years, it can easily be calculated (*See* problem 9) that the average power of the tides is 1,570,000 Mw. Actually, this calculation does not take into account several very major factors. While the earth is being slowed, the angular velocity of the moon moving around the earth has also been being slowed. In addition, some energy of the tides is also lost as low grade heat in the oceans. The actual power in the tides is far higher than that indicated above. It is obvious that there is considerable power in the tides; the problem is to harness this power.

Of course, only a small fraction of the existing tidal power can be utilized. In general, such utilization is only possible in selected geographic areas because the only practical, large-scale method of harnessing tidal power currently available is based on the use of one or more tidal basins. These basins must be capable of being isolated from the sea by dams or barrages. Power is generated by the use of hydraulic turbines with water flowing through these turbines either during the rising or the falling tides or during the transfer of water to or from one or more basins.

Only two tidal power stations are currently in operation. One at La Rance, France, with a capacity of 240,000 kw and a smaller one at Kislaya Guba, in the USSR, 600 miles north of Murmansk, with a capacity of 2000 kw. Figure 89 is an aerial view of the La Rance in-

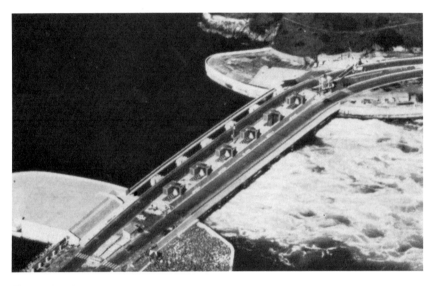

Figure 89. An aerial view of the La Rance tidal power station on the east coast of France. (Courtesy of the Office of the French Scientific Mission.)

stallation, which was dedicated in 1966. As may be imagined, these tidal power stations are in many ways similar to hydroelectric stations. In some important respects, however, tidal power stations are unique. In current hydroelectric stations, the turbines are designed to operate using a relatively high head of water. That is, the impounded water falls through a substantial pressure difference in the turbines. Tidal stations, however, use a very much lower head of water, and hence a much smaller pressure difference. The La Rance plant, for example, has a maximum head of 27 ft, while the Kislaya Bay station has a maximum head of only 11 ft. Thus, to achieve a large power output, power turbines have to handle extremely large water flows, when compared to the volume of water needed to produce a given amount of power in a typical hydroelectric station. Many years of design efforts have gone into hydroelectric turbine design, but only relatively recently has renewed engineering effort been devoted to the design of turbines for tidal power application. The La Rance turbines, for example, are of the so-called bulb type. They have a horizontal-axis propeller with variable pitch blades. The turbine is connected to a generator in an enclosed nacelle or bulb in the water passage for the turbine, as shown in Figure 90. They can be run either as power generating turbines or in reverse as pumps. This latter ability is especially useful in that this system can be operated as a pumped storage facility to use effectively off-peak power from associated standard thermal power stations. To store power, such reversible pump generators may be used to overempty the tidal basin at low tide or to overfill the basin at high tide. Since tidal power stations will involve salt water, there will be greater material corrosion problems than those encountered in fresh water hydroelectric power stations. Such corrosion problems have, in fact, been encountered at La Rance.

To estimate the amount of energy that may be available from a given tidal project, consider the energy stored by the water held to a depth, H, behind a dam of height, H_o. Let the surface area of this body of water be A and assume that the area doesn't change as the

Figure 90. Photographs of a bulb unit turbine at La Rance, seen from the basin side (left) and the sea side (right) (Courtesy of the Office of the French Scientific Mission.)

water level drops. This latter assumption is usually a reasonable approximation in many practical cases. The total weight of water stored, W, is then given if the dam is completely filled by

$$W = A \, H_o \rho \tag{87}$$

where ρ is the density of water (lb/ft³). As the water runs out of the reservoir (or tidal basin) the height, H, of the water drops, of course. The amount of energy released, dE, per unit of decrease in the height of water stored is then

$$dE = A\rho H dH. \tag{88}$$

To find the total energy stored, Equation (88) may be integrated between the limits of empty (H = 0) and full (H = H$_o$) to give

$$\int_{E=o}^{E=E_{Total}} dE \qquad \int_{H=o}^{H=H_o} A\rho H dH \tag{89}$$

or

$$E_{Total} = \frac{A\rho H_o^2}{2} \tag{90}$$

The total energy available therefore depends on the *square* of the height of the dam. In a similar way, the total energy available from a tidal energy scheme will depend on the square of the tidal range (difference between low and high tides), usually called R. To compute the maximum possible average power available from a given tidal scheme we must divide the total energy available by the total cycle time between high and low tides, T. In most cases T would be about 6 hours or 21,600 seconds. Thus, the average power that could be produced is given by

$$P_{ave} = \frac{A\rho R^2}{2T} \tag{91}$$

To give an idea of the magnitudes involved, consider a tidal basin with an area of 20 square miles and a tidal range of 8.5 ft. With T = 6 hours one then has, from Equation (91) with ρ = 62.4 lb/ft³.

$$P_{ave} = \frac{A\rho R^2}{2T}$$

$$= \frac{(8.5\text{ft}^2)\ (20 \times 5280^2\text{ft}^2)\ 62.4\text{ lb/ft}^3}{2(21,600)}$$

$$= 115,800,000\ \frac{\text{ft-lb}}{\text{sec}}$$

$$= 209,000\text{ hp}$$

$$= 159,000\text{ kw}$$

This is enough power to supply a small city, and confirms the intuitive notion that there is considerable power lost by not harnessing the tides.

We have been considering the most simple kind of tidal power system. Actually, such a one basin scheme would probably not be used in practice because the power available would vary greatly during the tide cycle. That is, the effective height through which the water would drop in passing through the turbines would be considerably less than the tidal range, R. With two basins, a low pool and a high pool, the power output could be made much more uniform. Consider the following: Allow the low pool to empty with the outgoing low tide and then be sealed. Allow the high pool to fill with the next incoming tide while the low pool remains empty. When this high tide starts to ebb, the high pool is isolated from the sea and allowed to empty through the turbines into the low pool. This process continues until the level in the low pool is equal to that of the decreasing level of the sea. The low pool is then opened and its level falls with the outgoing tide. In this way, the power output is made more uniform over a longer period than that which would be available from a single pool system. Without even more pools, however, there will still be periods of relatively low power production when the high pool is refilling.

Unlike wind power, tidal power installations when finally built, will tend to be on a large scale, comparable to those of current hydroelectric stations. As an example, there are many different dam configurations which have been proposed for a tidal power project in the Bay of Fundy (northeast of the Passamaquaddy region and substantially Canadian). These configurations range in size from 800 Mw (comparable to a medium-sized coal-fired plant) to 14,000 Mw (this would be approximately equal to New England's 1973 electric power requirement). Figure 91 shows a map of the principal sections of the Passamaquaddy tidal power scheme. The costs of tidal power stations in $/kw can be expected to vary widely and depend greatly on the geography of the bays and tidal areas involved. For the Bay of Fundy project, these costs were estimated at $475/kw (in 1967). This is not a very high number and is approximately equal to that of nuclear power stations. Clearly, tidal power will not solve all of our national energy problems. It does seem that tidal power, like hydroelectric power, can contribute a reliable, inexhaustible, power source, not subject to embargo, that should not be overlooked in the future as it has in the past.

In addition to the power contained in tidal water movement, the oceans also have power in the form of ocean waves. Substantially all ocean waves are the result of the interaction of wind and water. Wind blowing over calm water will cause ripples. Each such ripple is a slight impediment to the next air current, with the pressure on the

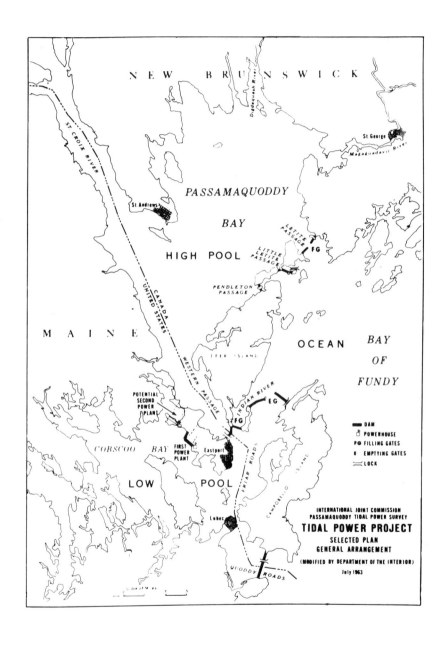

Figure 91. Schematic plan of the Passamaquaddy project.

windward side of the ripple being greater than the pressure on the lee side. Thus, the ripple will increase in size as a function of both wind velocity and direction. As an example, winds of 40 knots (40 nautical miles per hour or about 46 statute miles per hour) will produce waves 14 ft in height after blowing over water for six hours and these waves will increase to 35 ft if the wind continues blowing for 48 hours. In lakes, of course, large waves cannot form since the wavelets reach the shore. Hence, as we have seen, the winds are produced by solar energy, the waves are also indirectly dependent on solar energy. Other waves, such as so-called tidal waves (which have no relation to tides) can be caused by undersea earthquakes, but these waves are special cases.

Over the years there have been many schemes for harnessing the energy of waves. In 1909, for example, there existed an organization called the California Wave Motor Company which constructed small electric generating systems powered by panels caused to oscillate by wave action. These devices worked but were quite impractical for generating large amounts of energy. During his retirement to the Bras d'Or region of Nova Scotia, Alexander Graham Bell invented a device powered by wave action which utilized wave action to condense fog to form fresh water. This device, too, was never widely used. More recently, interest in harnessing waves has again revived, and many schemes have been proposed. The most practical appear to be those which hydraulically use wave action to produce an elevated reservoir of water. This water, in flowing back to ocean level can be used to drive a generator. Such systems have the advantage that no complicated mechanical assembly is in contact with waves. This is an advantage because, of course, waves can vary tremendously and any complicated system involving a large number of linkages is susceptible to failure in heavy wave conditions.

The ultimate use of ocean waves to produce usable energy is much less certain than that of tidal energy.

GEOTHERMAL ENERGY

The natural heat of the earth is thought to be produced by the slow decay of radioactive materials deep within the core of our planet. These nuclear processes have been going on for millions of years and will continue for millions more. It is believed that frictional forces resulting from solar and lunar tides, the relative motion of crustal plates, and the heat due to compression of subsurface material contribute to geothermal energy also, but this contribution is probably negligible compared to that produced by radioactivity.

Temperature measurements in drill holes, mines, etc., confirm that temperatures increase with increasing depth. The rate at which the temperature increases with depth (called the geothermal gradient) is very gradual. The distance from the surface to zones hot enough to be of value to us as sources of energy varies widely. However, "normal" temperature gradients range from 8°–15°C/km.

Theoretically, geothermal energy sources can be tapped from any point on earth simply by drilling deep enough holes, providing a passage for a heat transfer fluid, and extracting the heat. Practically speaking, however, present drilling technology is limited to about 12,000 meters, while the base of the continental crust, where the temperatures reach 1000°C, varies from 25–50 km in depth. At reasonable working depths, most of the heat in the geothermal resource base is too diffuse to be of economic value. Fortunately, sufficient concentrations of geothermal heat exist that we can utilize this immense resource base. In some regions, such as the vicinity of active volcanoes, geyser fields, and hot springs, the distance from the surface to hot rock can be a matter of only a few hundred meters.

Characteristics of Geothermal Sources

Before we examine the physical characteristics of geothermal energy sources it is instructive to consider briefly a cross section of the earth itself. The earth's crust varies in thickness from about 9.5 km of basaltic rock under the oceans to an average of 22.5 km of granite plus 9.5 km of basalt under the continents. The crust may be as thick as 48–65 km under mountains. Of course there are layers of soil, sand, and rocks such as limestone, sandstone, and many others. The mantle, believed to consist principally of high density rock, extends down 2900 km below the crust. The core of the earth is believed to be composed principally of 90% iron and 10% nickel with the outer 2100 km in a molten state (evidence based on studies of earthquake waves) and the inner core of 1350 km radius solid.

The earth's crust is divided into about six major plates which are in motion with respect to each other. The major known geothermal energy systems of the world are associated with the plate boundaries and other related geologic features. Figure 92 shows both the major geothermal energy systems and the geologic conditions which give rise to them. The double lines represent a worldwide system of ridges along which new ocean bottom is continually being extruded. The broken lines parallel to the ridge system represent the present position of material extruded at intervals of 10 million years, as determined by magnetic studies. In addition to representing plate boundaries, it should be noted that the barbed lines also follow some of the deep ocean trenches. The major earthquake activity in the world occurs along the various features shown in Figure 92.

Temperatures below the earth's surface are controlled principally by conductive flow of heat through solid rock, by convective flow in

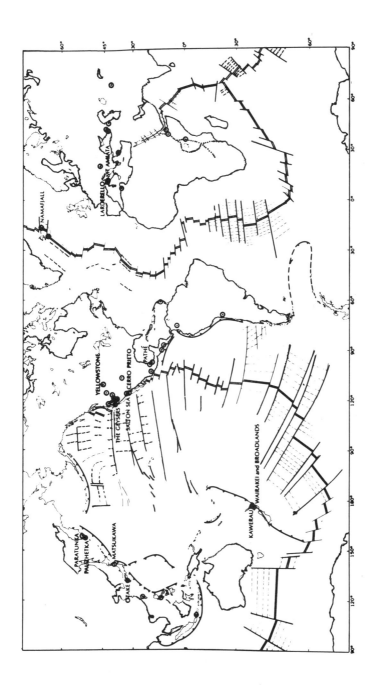

Figure 92. The major known geothermal systems of the world.[1] Named systems (other than Yellowstone) are those that are presently generating electricity or have power plants under construction. (Courtesy of Stanford University Press.)

circulating fluid fields characterized by extensive subsurface fracturing, and by mass transfer in magma, molten rock material in the earth's crust.

Conduction is the dominant mode of heat flow in most of the outer crust of the earth. The important relationship between thermal gradient, heat flow, and thermal conductivity is

$$q = K\Delta T$$

where heat flow (q) is expressed in μcal/cm^{2}sec, the thermal gradient (ΔT) is in °C/km, and thermal conductivity (K) is in mcal/cm sec°C. The worldwide average heat flow is about 1.5μcal/cm^2 sec, or 1.5 hfu (geothermal heat flow units). Heat flows ranging from 0.8–2.0 hfu are considered "normal." The variation is due to differences in the thermal conductivity of rocks which varies greatly with depth as a function of mineralogy, porosity, and fluid content of the pores. The thermal conductivity of most rock ranges from 4–10 mcal/cm sec°C. Generally, the heat transferred by conduction is thought to be too diffuse to be considered for possible exploitation as a geothermal energy source.

The source of this heat is believed to be the small amounts of radioactive elements contained in most rocks. The isotopes which give off important amounts of heat are given in Table 24. The two principal rocks found in the earth's crust, granite and basalt, contain the following amounts of these isotopes:

Granite	4.7 ppm U	20 ppm Th	3.4%K
Basalt	0.66 ppm U	2.7 ppm Th	0.8%K

A calculation of the amount of heat produced by this radioactivity shows that the observed average heat flow is easily accounted for. (Problem 1).

Table 24. Heat production by long-lived radioactive isotopes and their products.[a]

Isotope	Half Life	Proportion of Isotope	Heat generation cal/gram/yr
U^{238}	4.5 x 10^9 yr	99.28%	0.70
U^{235}	0.7 x 10^9	0.714%	0.03
Th^{232}	13.9 x 10^9	100.0%	0.20
K^{40}	1.31 x 10^9	0.12%	27 x 10^{-6}

[a]"Geothermal Energy," The Unesco Press.

By contrast to conductive thermal gradient regions, there are hydrothermal-convection systems in which most of the heat is transferred by circulating fluids rather than by conduction. Two major types of hydrothermal convection systems are recognized and, due to the relatively high-energy density in these systems, both may be used as geothermal energy sources. Both *hot-water* and *vapor-dominated* systems are characterized by the features shown in Figure 93. The principal features are a) a large source of natural heat such as a magnetic intrusion, b) an adequate water supply, c) an aquifer, or permeable rock, and d) a cap rock. Water serves as the medium by which heat is transferred from deep sources to a geothermal reservoir at shallower depths. Cool rain water, falling in areas ranging from tens to thousands of square kilometers, circulates downward. At depths ranging up to a few kilometers, the water is heated by conduction from hot rocks, expands upon heating, and then moves buoyantly upward in a column of relatively restricted cross-sectional area (1–50 km^2). Whether or not this geothermal energy is stored or escapes to the surface depends upon the degree of interconnectivity of pores and fractures in the cap rocks.

A few geothermal systems, including the Larderello Fields of Italy and the Geysers of California, produce dry or superheated steam with no associated liquid. *Vapor-dominated systems* such as these are believed to have developed thousands to tens of thousands of years ago. They probably consisted initially of hot-water systems characterized by a very high heat supply and very low rates of recharge

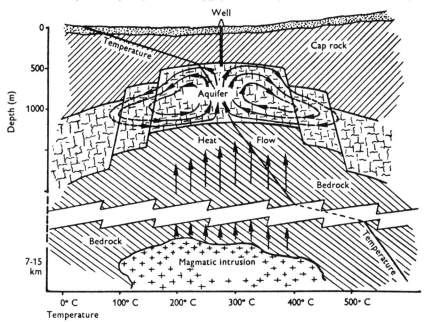

Figure 93. Basic model of a geothermal steam field. (Source: *Geothermal Energy*, The Unesco Press.)

(return of surface or pore water to replenish that driven off as vapor). When the heat supply of a developing system becomes great enough to boil off more water than is being replaced by recharge, a vapor-dominated reservoir begins to form.

Utilizing Geothermal Systems

There are many problems associated with large-scale utilization of geothermal energy. In about 80% of all hydrothermal reservoirs, the temperature is not high enough to provide fluids that may be used directly to drive a turbine. In addition to insufficient temperatures, dissolved gases and salts, and particulate matter contained in the steam and hot water present serious operating problems.

Gases such as CO_2, H_2S, and NH_3 are both volatile and soluble in water. Their removal is necessary to maximize plant efficiency, to minimize corrosion of plant equipment, and to offset adverse ecological effects. Dissolved salts such as silicates can cause fouling of turbines and decrease their performance. They also deposit and restrict the flow in pipes. Boron can be detrimental to plant life. Particulates in steam cause erosion of turbine blading and valves. To minimize erosion and to maintain good performance from transfer equipment, as much particulate matter as possible should be removed from the steam before it reaches the utilization plant.

Another problem is controlling the considerable noise generated by the release of steam during well venting, during release of over-pressures in the steam collection system, and during general in-plant operations. Technology currently exists for minimizing this problem.

To eliminate the disadvantages inherent in utilizing the contaminated geothermal fluids discussed above, vapor-turbine cycle systems, which operate on the same principle as a heat pump or refrigerator, have been developed. In these systems, the hot geothermal water and steam are passed through a heat exchanger causing a working fluid such as freon or isobutane to boil and superheat. This vapor is expanded through a turbine to produce power and is then recirculated to the heat exchanger. The cooled water may then be recharged into the ground depending on the characteristics of the geothermal source.

Current Status of Geothermal Energy

Current geothermal electric power stations are shown on the map in Figure 94. The electric power capacity of several of these stations, as well as projected development is shown in Table 25.

The oldest commercial-sized installation to generate electricity from natural steam is the Larderello development in Italy. In 1913 a 250-kw generator was installed. Since that time, the Larderello field

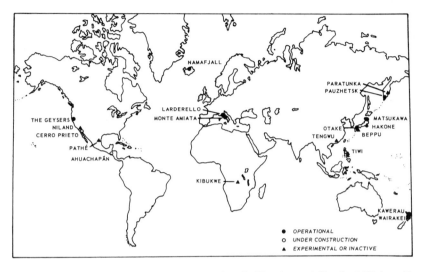

Figure 94. Geothermal electric power stations.[2] (Courtesy of Stanford University Press.)

Table 25. Expected Development of Electrical Power Capacity at Selected Geothermal Fields to 1980.[2]

Nation	Field	Installed Capacity, late 1972, (Mw)	Expected Development (Mw)
El Salvador	Ahuachapan	—	30 by 1975; 60 by 1980
Iceland	Hengill(Hveragerdi)	—	Up to 32 by 1980
	Namafjall	3	None known
Italy	Larderello	365	15% increase possible
	Monte Amiata	25	
Japan	Hachimantai	—	Perhaps 10 by mid-1970s
	Matsukawa/Takinokami	20	Perhaps 60 by 1980
	Onikobe	—	Perhaps 10 by 1980
	Shikabe	—	7; salt-recovery works planned for 1970s
	Otake/Hachobaru	13	Perhaps 60 by 1980
Mexico	Cerro Prieto	75	150 by 1980
New Zealand	Kawerau	10	None planned
	Wairakei	160	None planned
U.S.S.R.	Pauzhetsk	6	Up to 25 by 1980
	Kunashir	—	Up to 13 by 1980
United States	The Geysers	302	110/year through 1980, to 1180
	Imperial Valley	—	Demonstration for desalination; power station by 1980

and its neighboring steam producing area, Monte Amiata, have grown to 390-Mwe capacity. It is currently the world's largest.

The most developed utilization of geothermal water is in Iceland. Hot water for municipal heating was first used in the 1930s. Approximately 50% of the 200,000 population receives geothermal heating, and this is expected to rise to 60% by 1980. Nine out of ten homes in Reykjavik, the nation's capital, receive geothermal water for home heating. Geothermal electric power generation was established relatively recently when a 3 Mwe plant was opened in 1969 at·Namafjall. Due to the wide distribution of cheap and abundant heat energy, multiple, small-capacity power plants appear to be the most effective scheme for its utilization.

The Wairakei and Kawerau geothermal fields in New Zealand are part of a dozen or more major geothermal areas in the Taupo Volcanic Zone. This volcanic zone is associated with the Tonga deep ocean trench and major geologic faults lying along the western edge of the Pacific crustal plate. About 15% of the electrical power generated in New Zealand is derived from geothermal sources. These sources are hot water sources similar to those found in Iceland. A small fraction of the hot, high-pressure water flashes to steam when it is exposed to near atmospheric pressures near the surface.

The principal geothermal source operating in the United States is at the Geysers, California. This area is one of the two major dry-steam areas being utilized, the other being the Larderello fields in Italy. The first turbine of 12,500 kw was installed in 1960. In 1972, 302 Mw of electricity were produced. Another 110 Mw will be installed annually through the 1970s, and by 1980 a total capacity of 1180 Mw will exist at the Geysers. Figure 95 shows the Geysers geothermal steam field in California. Numerous steam vents can be seen in the photograph.

Of the other potential geothermal sites in the United States, the most significant are in the Imperial Valley of California. A plant at Niland in the Imperial Valley was expected to be in operation in 1974. Space heating with geothermal fluids is succeeding in southcentral Oregon. At Klamath Falls some 350 wells supply heat to buildings via heat exchanging with pure municipal water. Elsewhere in Oregon, greenhouses, resorts, baths, farm buildings, and schools are heated by geothermal waters. Similar activities take place on a more limited scale or are proposed in Idaho, Utah, New Mexico, Arizona, and Nevada. A recent extensive survey of the geothermal potential in Montana showed gradients as high as 75°C/km. Utilization of these sources will parallel improving technology for handling hot water-wet steam systems.

The geothermal potential at Yellowstone National Park was examined by the U.S. Geological Survey from 1967–1969. It is the world's

Figure 95. Geysers geothermal steam field.

most extensive display of hot spring and geyser phenomena, but it is permanently set aside as a National Park and no development is anticipated.

Summary

What are the prospects for geothermal power? The answer to this question requires ascertaining the technical and economic feasibility and environmental acceptability of geothermal resources development. Most of the current developments center around dry-steam and hot-water systems. Economically competitive power is being generated in these areas of relatively high-energy density. However, most of the enormous heat content of the earth's interior that is within reach by today's drilling technology is present in the dry rock. And most of this energy is very diffuse; it is so-called "low-grade" energy. Schemes to concentrate this energy, such as by underground nuclear detonation to produce fracture patterns for hydrothermal-convection systems are being considered.

OCEAN THERMAL ENERGY

The sea covers some 140 million square miles or 71% of the earth's surface. The average depth of the sea is 2.5 miles. Hence, the total volume of water in the ocean is roughly 350 million cubic miles. These oceans, particularly the tropical seas, are built-in collectors of the sun's heat. For example, for every Fahrenheit degree temperature rise, a cubic mile of sea water absorbs about 2.69×10^9 kwh of energy. However, like the natural heat of the earth, this energy is very diffuse — of "low grade."

When two extensive currents of water, one warm and one cold, exist in close proximity to one another, it is possible to operate a power plant utilizing this low-grade energy. There are many locations in the world where vast ocean currents run within 2000–3000 ft of each other. The major surface currents found in the earth's ocean are shown in Figure 96. The speeds of the ocean currents are generally slow, less than 1 knot (1 nautical mile per hour). However, in a few places such as the Gulf Stream, Kuroshio, and the Equatorial currents, speeds greater than 1 knot are usually encountered and speeds of 5 knots are sometimes observed. The surface currents in the oceans are, with respect to the total depth of the ocean, relatively shallow features. In general, the movements extend to a depth of several hundred feet, while in areas of high speed currents, such as the Gulf Stream, they may extend as far as 3300 ft below the surface.

The temperature of the surface water depends on the latitude; average temperatures are shown in Table 26.

Table 26. Average Surface Temperatures of the Oceans.[3]

Latitude	90°N	80°	70°	60°	50°	40°	30°	20°	
Average Temperature	29°F	29°	34°	41°	46°	59°	71°	78°	
Latitude	10°	0°	10°S	20°	30°	40°	50°	60°	70°
Average Temperature	82°	81°	79°	75°	68°	57°	42°	32°	30°

At a depth of about 3300 ft below the surface, temperatures range from 36°–41°F. The cold water flows along the ocean floor from the poles toward the tropics. It warms up in the tropics, rises to the surface, and flows back again along the surface.

Unlike tidal and wind energy, the thermal energy stored in the sea is available continuously. Theoretically, this energy may be extracted wherever a temperature difference exists. Practically, however, the extraction of such energy becomes more difficult, more costly, and less efficient the smaller the temperature difference between the high- and low-temperature reservoirs.

Figure 96. The actual system of ocean currents. (Courtesy of D. C. Heath and Company.)

Power Production

As far back as 1882, D'Arsonval suggested that power could be generated by using the temperature gradients found in the ocean. It was not until 1926, however, that Georges Claude, a French scientist and friend of D'Arsonval, built a land-based plant at Matanzas Bay, Cuba. Claude's design incorporated a low-pressure steam cycle. Warm surface water was pumped to an evacuated chamber where boiling took place due to the reduced pressure. The steam was passed through a turbine and condensed by direct contact with cold seawater pumped from the ocean bottom to a land-based pit. The plant finally produced power in 1929. It was designed for 40 kw and reached 22 kw before various problems forced shutdown. The chief problems were a) the use of water itself as the working fluid, and b) the extremely long lines needed to bring water from the depths. Heat flow into the long lines and corrosion of both the lines and turbine were excessive.

An alternative to Claude's land-based ocean thermal energy plant is to build a plant which floats in the warm water stream. Figure 97 shows a huge, 400 Mwe ocean thermal-difference generator,[4] which is a concept developed by the Energy Research Team of the University of Massachusetts headed by Professor William E. Heronemus. The generator plant would be tethered in the Gulf Stream by a single-point mooring and would ride there like a kite. This concept uses a closed thermodynamic cycle in which warm surface water and cold subsurface water pass through heat exchangers and respectively evaporate and condense a working fluid such as propane, ammonia, or one of several fluorocarbons such as freon. In the drawing, cold water is shown being sucked up from about 75 ft above the sea floor through an inlet pipe (1), passing through inlet doors (2), into condensers (3) inside twin 85-foot diameter hulls, and discharged through outlets in the walls of the hulls. The cold water pumps (4) maintain this circulation. At the surface, warm water is sucked through pod mounted evaporators (5) by warm-water circulating pumps (6). Ballast tanks (7) maintain a negative buoyancy so that the entire structure floats just beneath the ocean surface.

The working fluid flows from a pressure reservoir (8) to an evaporator feed pump (9), which pressurizes it, and on through the evaporators (5) where it is boiled into a vapor by the energy extracted from the warm seawater. The vapor passes through turbines (10) which, in turn, drive electrical generators (11). The vapor then expands through turbine exhaust diffusers (12) and through the condensors (3) where its residual heat is extracted by cold water. The condensed liquid then returns to the reservoir (8) and begins the cycle again. Each hull contains eight power packages consisting of a condensor, turbine, and generator. The hulls are pressurized at

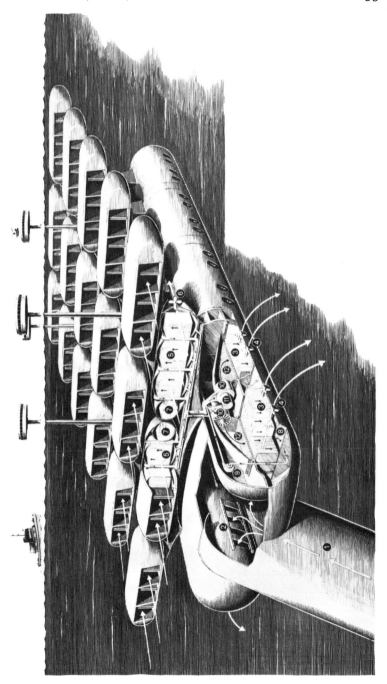

Figure 97. 400-Mw ocean thermal-difference generator (Reproduced from *Popular Science* with permission Times Mirror Magazines, Inc. and Ray Pioch, artist.)

one atmosphere and each hull contains eight power packages that generate 25 Mwe electrical power.

The maximum theoretical thermodynamic efficiency of an ocean thermal-difference power plant is about 6.5%, but various inefficiencies and losses would reduce the actual efficiency to something less than 2.5%. Thus, to obtain useful amounts of energy, enormous amounts of warm water would have to pass through such a plant.

All of the technology for an ocean thermal difference process exists. Capital investment costs are quite high, however. They are estimated to be roughly $550–600/kwe. These costs include an umbilical for transmitting the electrical energy to shore. By comparison, capital investment costs for fossil fuel and nuclear power plants are in the range $200–300/kwe. Of course, the overall economics of the ocean thermal facility compare more favorably when the "free" fuel is figured in for a long-term operation.

Summary

The future of ocean thermal energy is unclear. Certainly there is a vast supply of renewable energy available. One concept for utilizing this energy, the ocean thermal-difference concept, has been discussed. This system is based essentially on state-of-the-art technology. To what extent ocean thermal energy is utilized in the future depends to a large extent on whether such energy can be obtained economically.

REFERENCES

1. White, D. E. "Characteristics of Geothermal Resources," in *Geothermal Energy,* P. Kruger and C. Otte, Eds. (Stanford, California: Stanford University Press, 1973), p. 70.
2. Koenig, J. B. "Worldwide Status of Geothermal Resources Development," in *Geothermal Energy,* P. Kruger and C. Otte, Eds. (Stanford, California: Stanford University Press, 1973), p. 20.
3. Wolfe, C. W., L. J. Battan, R. H. Fleming, G. S. Hawkins and H. Skornik. *Earth and Space Science,* (Boston, Massachusetts: D. C. Heath and Company, 1966), p. 464.
4. Fisher, A. "Energy from the Sea: Part II: Tapping the reservoir of solar heat," *Popular Science* p 78-79, (June, 1975).

SUGGESTED FURTHER READING

Windpower

1. Savino, J. M. (Ed.), "Wind Energy Conversion Systems," Report No. NSF/RA/N-73-006 (Springfield, Virginia: National Technical Information Service, 1973).

2. Putnam, P. C. *Power from the Wind,* (New York, New York: Van Nostrand Reinhold Co., 1948).
3. Golding, E. W. *The Generation of Electricity by Wind Power,* (London: E. & F. N. Spon, Ltd., 1955).
4. Vargo, D. J. "Wind Energy Developments in the 20th Century," NASA Technical Data Memorandum No. NASA TM X-71634. (Springfield, Virginia: National Technical Information Service).
5. Savino, J. M. "A Brief Summary of the Attempts to Develop Large Wind-Electric Generating Systems in the U.S.," NASA Technical Data Memorandum No. NASA TM X-71605 (Springfield, Virginia: National Technical Information Service).

Tidal Power

A full listing of tidal power publications is available from the Reference Section, Science and Technology Division, Library of Congress, 10 First Street, S.E., Washington, D.C. 20540. Request the LC Science Trace Bulletin on Tidal Power, compiled by Jane Collins.

1. Gray, T. J. and O. K. Gashus (Eds.) *Tidal Power Proceedings,* International Conference on the Utilization of Tidal Power, Nova Scotia Technical College, 1970, (New York, New York: Plenum Press, 1972) pp. 630.
2. Davey, N. *Studies in Tidal Power,* (London: Constable and Co., 1923) pp. 255.
3. Turnbull, W. R. "Proposed Tidal Hydroelectric Power Development of the Peticodiac and Memramcook Rivers," in *Annual Report of the Smithsonian Institution,* 523-546 (1923).
4. Macmillan, D. H. *Tides* (New York, New York: American Elsevier Publishing Co., 1966) pp. 240.
5. Shaw, T. L. "Tidal Energy and National Needs," *Water Power,* 24, No. 10, 324-384 (1972).

Geothermal Energy

1. Armstead, Christopher H. (Ed.) *Geothermal Energy,* (Paris France: The Unesco Press, 1973).

Ocean Thermal Energy

1. Anderson, J. Hilbert and James H. Anderson, Jr. "Thermal Power from Seawater," *Mech. Eng.* 88, No. 4, 41-46 (April, 1966).
2. Anderson, J. Hilbert and J. H. Anderson, Jr. "Power from the Sun by Way of the Sea?," *Power,* 64 (January, 1965); 63 (February, 1965).
3. Claude, Georges. "Power from the Tropical Seas," *Mech. Eng.* 52, No. 12, 1039–1044 (1930).
4. Gerard, R. D. and O. A. Roels. "Deep Ocean Water as a Resource," *Marine Technol. Sci. J.* 4, No. 5, 69 (October, 1970).

5. Green, Jack. "A Self-Contained Oceanic Resources Base," *Marine Technol. Sci. J.* 4, No. 5, 88 (October, 1970).
6. Waters, Samuel. "Power in the Year 2001," *Mech. Eng.* 93, No. 10, 21 (October, 1971).

PROBLEMS

1. The specific density (gram/cm^3) of air at 15°C and a barometric pressure of 760 mm of Hg is 0.00122. Calculate the power, in kilowatts, of a 10-mph breeze passing through the area swept out by a wind turbine 131 ft in diameter.
2. A flapping flag also extracts energy from the wind. Try to design a practical method for extracting useful work from a flapping flag.
3. An isolated farmhouse is to be powered by a windmill. The house needs 2000 watts. The wind blows at a steady 13 mph. If the windmill generator combination can capture 10% of the power in the wind and turn this into electricity, what diameter (in feet) must the windmill blades be. Assume the air in the region of the farmhouse weighs 0.087 lb/ft^3.
4. Would you expect the power per unit area of a 10-mph breeze in the desert to be greater, less than, or equal to the power per unit area of a 10-mph breeze from off the ocean? Explain your answer. Hint: Compare the densities of wet and dry air. Is this a significant effect?
5. How many wind turbines having blades 100 ft in diameter and situated in a region where the wind velocity was 15 mph would it take to supply the 15,000 Mw of electricity required by New England. Assume these wind turbines have the maximum possible theoretical efficiency and that none of the mechanical power produced by the wind is lost during conversion to electric power.
6. Normal turbines have precise speed controls so that the frequency of the electrical power that they produce is always 60 cycles per second. The rotational velocity of wind turbines will not be constant and hence, the frequency of the electricity that they produce will also not be constant. Discuss at least two possible ways of dealing with this problem.
7. At the Balaclava site of the Russian wind machine discussed at the beginning of the chapter, the average year-round wind velocity is 13 mph. Based on the data given for this system, with what efficiency did it turn wind power to electrical power.
8. The NASA-Lewis 100-kw wind turbine discussed in the text will reach its rated output in an 18-mph wind. What is the efficiency of this machine?
9. Show that the value of 1,570,000 Mw given in the text derives from the proper use of Equation (82) together with the data given

plus the fact that the earth is 6371 km in diameter, has a mass of 5.98×10^{24} kg, and rotates once on its axis (goes through 2π radians) every 24 hours.

10. New England is very dependent on Arab oil for electrical power. The Passamaquaddy area in northern Maine has the highest tides in the world. How large an area (in square miles) would have to be used to save 100,000 barrels of oil per day if the average tidal rise and fall (high tide to low tide) were 30 feet? One barrel of oil can produce about 500 kwh. Take the density of seawater to be 64.16 lb/ft³.

11. Why do you suppose the Passamaquaddy area is still not developed? By 1935, $7 million had been spent on this project before the money to finish it was withdrawn by Washington. Was this a mistake?

12. Suppose a tidal basin has an area of 20 square miles and that the tidal range is 16 ft. After the basin is completely full, and the tide has receded to its low point, what energy is potentially available from the water impounded in the basin? Express your answer in both ft-lb and kwh.

13. One plan for Passamaquaddy involves a planned high water basin area of 101 square miles and an average tidal range of 18.1 ft. What is the maximum *average* power that could be produced by such a plan assuming that the average time between high water is 12 hours, 25 minutes. Take the density of seawater as 64.16 lb/ft³. Express your answer in megawatts. The actual expected average output of this system is 1000 Mw.

14. If the tides are primarily due to the rotation of the earth and the moon, how do you explain the fact that the average time between two high waters is, typically, about 12 hours?

15. a) Assuming a uniform heat flow of 1.5 μcal/cm²sec over the entire earth's surface during the past 1000 million years, what is the total heat flow in cal/cm²?

b) If this amount of energy was derived from the combustion of coal, determine how many tons of coal under each cm² of surface would be required. Assuming an average density for coal of 1.5 gram/cm³, determine the thickness of this hypothetical coal "vein."

16. a) Assume the volume of hot, dry rock associated with the volcanic area surrounding the Valles Caldera in the Jemez Mountains in northern New Mexico is 5600 km³. How many calories of heat would be released if the average temperature of this material were to decrease by 1°C. Assume the average rock density is 2.75 gram/cm³ and the heat capacity is 0.20 cal/°C/gram. (Hint: Energy = mass x heat capacity x ΔT).

b) The conversion efficiency for producing electrical energy from geothermal energy is about 14%. At this conversion efficiency, how many Megawatt-centuries of electrical energy does your answer in part (a) equal?

17. An ocean thermal difference plant is to be located in the Gulf Stream where the surface water temperature is 78°F and the subsurface water temperature is 36°F. The power rating of this plant is 400 Mwe and the thermal efficiency is 2.2%. If the temperature of the warm water falls 10°F during transfer of energy to a working fluid such as propane, how many ft^3/hr of warm water must pass through the plant? (Specific heat of water $= 1.0$ Btu/lbm°F, density $= 64$ lbm/ft^3).

Chapter 11

Fuels for the Future—Coal, Shale Oil, Hydrogen, Methane

INTRODUCTION

We have considered several nonfossil energy sources such as nuclear, solar, wind, tidal, and have discussed the primary fossil fuel resources—coal, oil, and natural gas. With the exception of coal, it has become apparent that our fossil fuel resources are quite limited in the long term.

Coal is in abundant supply in the United States. We have about one half the free world's known supply, representing over four times more energy than the Middle East has in oil. Our 3 trillion tons of coal of various grades underlie various amounts of overburden and certainly not all of this coal can be recovered economically. Modern mining methods could recover perhaps 2 trillion of the 3 trillion tons. At the present rate of energy consumption, this amount of coal is sufficient to meet all energy requirements for several centuries. Figure 98 shows the location of coal producing areas in the contiguous United States.

While the total quantity of coal consumed annually is increasing, its fraction of the total energy market has declined during the 20th century. This decline can be correlated with the increased usage of other fossil fuels, oil and natural gas, which contain much higher percentages of hydrogen. Increased utilization of coal and diversification of its applications to include such activities as transportation will require continuing our efforts to solve the pollution problems attending the use of high-sulfur coal, and continuing coal gasification and liquifaction research and development programs.

Another fossil fuel resource in great abundance in the United States is shale oil. Oil-bearing shale is a mixture of clay, sand, and limestone in varying proportions, and kerogen, a greyish-brown rubbery

231

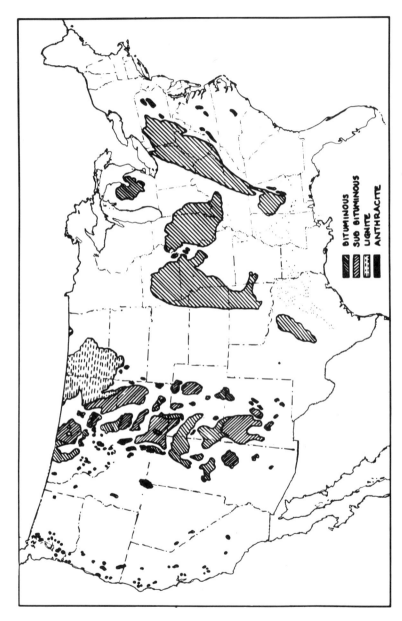

Figure 98. Location of coal-producing areas in the contiguous United States. (Courtesy of Energy Research and Development Administration.)

material. Kerogen is generally present in a ratio of one part to five parts of shale. Spread over some 11 million acres of Colorado, Utah, and Nevada there are an estimated 600 million barrels of economically recoverable shale oil. This is nearly equivalent to the known reserves of crude petroleum in the whole world. We will discuss briefly the technology of shale oil recovery later in this chapter.

Many of the nonfossil energy sources, while they are "renewable," are limited in that their availability is periodic (*e.g.*, wind, tides), they are available only in specific geographic areas (geothermal, tidal, ocean thermal), or, like nuclear energy, they are most efficiently utilized in large stationary operations. Also, energy demand cycles often do not coincide with production cycles, and "load centers" may be remote to the energy sources. While the future for some of these energy sources is unclear at the present time, the technology for harnessing such energy forms either exists or is presently under development. However, as these new energy sources are phased in, increased provisions for energy transmission and storage will be required.

One possible solution to the intermittant supply and geographic location problems of the energy sources mentioned above is to utilize those sources of energy to produce a synthetic fuel which is easily transported and stored. Hydrogen produced by the electrolysis of water is such a possibility and will be discussed.

Methane, the principal component of natural gas, produced by treatment of cellulosic solid wastes is another synthetic fuel under consideration. While the potential amount of such methane is much less than the other synthetic fuels cited, it is nevertheless significant and will be discussed also.

COAL GASIFICATION

Hydrocarbon fuels such as oil and natural gas are generally considered to be more versatile than coal. This versatility is related to the hydrogen content of the fuel, with oil and natural gas exhibiting higher hydrogen contents, 13 and 25 wt %, respectively, than coal (~7.0 w/o). In addition to determining its state at ambient temperatures, the hydrogen content determines such physical properties as the boiling point, viscosity, etc. More importantly, the heating value per unit of these fuels increases with hydrogen content. (*See* Appendix 3).

Since coal is relatively low in hydrogen, the primary objective of all gasification and liquefaction processes is the addition of hydrogen. There are several processes which accomplish this. The principal difference between them is in the operating conditions; differences in chemical reactions are secondary.

An overall general gasification flow sheet is shown in Figure 99. It should be noted that the gasification of the coal takes place in Stage 1 only. The remaining stages represent processes for removal of CO_2, and H_2O which have no heating value, removal of sulfur, generally as H_2S, and upgrading the heating value of the gas by methanation.

Ideally, it would be desirable to gasify coal according to the reaction

$$2C + 2H_2O \rightarrow CH_4 + CO_2 \qquad \Delta H = 2,765 \text{ cal/gram mole} \qquad (92)$$

because methane (CH_4), the principal component in natural gas, is produced directly. Furthermore, the reaction is only slightly endothermic (indicated by the small positive ΔH) and, thus, would require only a small amount of heat to sustain itself. Other reactions compete with the above reaction, however, and since several of them are favored, methane production is indirect. That is, in addition to a small amount of CH_4 produced directly in the gasifier, the exit gases will contain CO, CO_2, H_2, H_2S, and H_2O as shown in Figure 99. NH_3 may be present also depending on whether air or oxygen is used in the process.

The water gas shift reaction, shown as Stage 2 in Figure 99, is

$$CO + H_2O \rightarrow CO_2 + H_2 \qquad (93)$$

The objective of this operation is to bring the mole ratio of H_2 to CO to a value of 3 to 1. With this ratio, CO can be converted to CH_4 in the presence of a suitable catalyst (Stage 4).

The acid gases, H_2S and CO_2 are removed in Stage 3. This is normally done in a wet collector called a scrubber. The most simple wet collectors are spray towers which promote contact between the gas and a scrubbing liquid by violent action within a narrow throat section into which the liquid is introduced. Proper choice of a scrubbing liquid allows selective absorption of H_2S and CO_2.

Methanation is accomplished in Stage 4 by the conversion of CO and H_2 according to the reaction

$$CO + 3H_2 \rightarrow CH_4 + H_2O \qquad \Delta H = -49,071 \text{ cal/gram mole} \qquad (94)$$

This reaction must be catalyzed to proceed at an economical rate. The large, negative ΔH indicates that the reaction is strongly exothermic and cooling must be provided.

The methane produced in Stage 4 is referred to as synthetic natural gas (SNG) and must be dried (Stage 5) to prevent water from robbing it of its heating value. Gases are usually dried by passing them through some agent such as sulfuric acid, calcium chloride, or silica gel. The water vapor is absorbed by these dehydrating agents, generally resulting in chemical combination between the water and the drying agent.

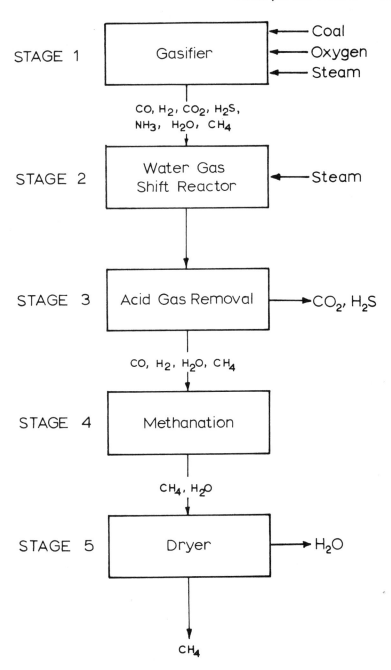

Figure 99. General gasification flow sheet (after Kermode.[1]).

Synthetic natural gas has the same heating value as natural gas, about 1000 Btu/ft^3, and is referred to as pipeline quality gas. Its heating value is such that it can be economically transported via existing pipelines.

Gas produced in the initial gasification process, Stage 1, is called producer or fuel gas. Its heating value is less than 250 Btu/ft^3, typically 135–165 Btu/ft^3. It requires cleaning, but then can be used for power generation by combustion in a gas turbine or in a boiler for steam production. Its heating value is low enough that it cannot be transported economically from the site at which it is produced.

Synthesis gas can be produced in Stage 2 by adjusting the ratio of H_2 to CO to 2 to 1. Such gas has been used for many years for making synthetic motor fuels and chemical compounds. Its heating value is in the range 250–400 Btu/ft^3.

Specific Processes

Several specific coal gasification processes have progressed significantly toward commercialization. Among them are the so-called BIGAS, HYGAS, CO_2 Acceptor, and Synthane processes. Pilot plants for each of these processes, with coal capacities ranging from 40–120 tons/day and gas production from 0.4–2.4 MMCF/day, either have been operated or were scheduled for operation by mid-1975. (MMCF is millions of standard cubic feet (SCF)—volume of gas at 60°F, 1 atm.) Differences between these processes include the type of coal used, mechanisms for feeding coal to the gasiffer, synthesis gas clean-up methods, and the methanation system used. Commercial plant projections for each of these processes are shown in Table 27.

A fifth coal gasification process which should be mentioned is the Lurgi process. This process was developed in Germany in the 1930s and is the only process commercially available at the present time. Several industrial plants are in operation in Europe and South Africa and a full-scale (250 MMCF/day) commercial plant is being built at

Table 27. Commercial Plant Cost Projections (Plant Production 250 x 10^6 ft^3 SNG/day).[a]

Gasification Process	Estimated cost, \$ x. 10^6	Coal Consumption	Product Price \$ per 1000 SCF
BIGAS	300	5 x 10^{6}T/yr	1.10
HYGAS	300	5 x 10^{6}T/yr	1.15
CO_2 Acceptor	150	30,000 T/day	.95–1.11
SYNTHANE	200	15,000 T/day	.80–1.00

[a]data from Office of Coal Research publication *Clean Energy from Coal Technology.*

Four Corners, New Mexico, for the El Paso Natural Gas Co. Although a large amount of commercial experience with the Lurgi process exists, it has several technical problems as well as a disadvantage of high capital costs.

Before discussing coal liquefaction, *in situ* gasification of underground coal should be mentioned. In-place combustion of coal is of interest as a possible way to extract the heat energy from coal that cannot be mined economically or from coal remaining after deep mining operations. Studies of *in situ* coal gasification processes have been conducted by the U.S. Bureau of Mines, the former Atomic Energy Commission, and others. The concept consists of several steps. First, chemical explosives are used in an array of drilled holes to fracture a coal bed creating passages for the flow of gasification fluids and the gaseous product. Fracturing could also be done by hydraulic pressurization. Collection wells would be drilled to the bottom of the fractured zone. The top would be ignited and a steam-oxygen mixture would be pumped into the coal vein. The product gases would contain CH_4, CO, CO_2, H_2S, and H_2 similar to surface gasification operations. This gas could be burned for electric power generation after cleaning or be further processed by methanation to upgrade its heating value.

Liquefaction

Several processes for converting coal to liquid and solid fuels also are currently under development. Among them are the COED (coal-oil-energy-development), COG (coal-oil-gas), H-Coal, and SRC (solvent refined coal) processes. Much of this work is supported by the U.S. Department of the Interior's Office of Coal Research. One of the principal products of coal liquefaction is called solvent refined coal (SRC) which is a heavy organic material having a melting point of 350°F and containing about 0.1% ash and less than 1% sulfur. Solvent-refined coal is produced by dissolving coal in a solvent under moderate hydrogen pressure. The solution is filtered to remove ash and undissolved matter. Finally, the solution is treated to remove the solvent. SRC has a heating value of about 16,000 Btu/lb, regardless of the quality of the original feed stock. Depending on how much solvent is left in the final product, SRC can be used in either liquid or solid form. As a low-sulfur, low-ash solid it can be used in power plants and industrial installations that now use coal. In liquid form, it can be hydrogenated by one of several processes under development to yield low sulfur fuel oil or synthetic crude oil (syncrude) for conversion to gasoline and other petrochemicals.

Numerous economic and resource evaluations of the potential for extensive coal gasification and liquefaction have been made. Synthetic fuel production is a highly capital intensive industry. Projected capital costs for 250 MMCF/day gasification plants and 40,000 bbl/day syncrude plants range from $350–$500 million per plant, and this cost is escalating. Many such plants would be required to contribute significantly toward meeting future fuel requirements. For example, it is estimated that 100 syncrude plants would be required to produce only one-fourth of the 15 million barrels of oil consumed daily in 1974. In addition to capital costs, an enormous demand on coal and land reserves, water reserves, skilled manpower, materials, and construction capacity would have to be met. Nevertheless, converted coal appears to be a likely fuel of the near future. Research and development of many conversion processes is continuing and, in all cases, operability and reliability will play a major role in determining the success of this venture.

SHALE OIL

Conceptually, the recovery of oil from shale is straightforward. Several processes are under development but the one most likely to reach commercialization first is the TOSCO (The Oil Shale Co.) process. This process is depicted in Figure 100. Mined rock is crushed to one half-inch pebbles and fed to a preheat tower where it is heated to about 500°F by hot gas. This material is then fed into a long, slanted, rotating drum, where it mixes with marble-sized ceramic balls which have been preheated to 1600°F. The hot ceramic balls pulverize and heat the shale until oil and gas boil off. The spent shale is screened out and the oily vapors are fractionated as shown.

Other processes under development utilize product gas to provide the energy required to heat the shale. Typically mined shale is fed into the top of a cylindrical retort and is fed by gravity toward a combustion zone. Such retorts are similar in principle to a blast furnace used to make pig iron.

One of the biggest problems associated with shale oil extraction is that of disposing of the spent rock. Since the rock is crushed up during processing the void space between rock fragments results in an overall expansion of about 12%. Thus, 12% (by volume) more material must be returned to the site than was originally mined. This presents an ecological problem which must be solved and is being intensively studied.

One scheme for possibly eliminating the dumping problem is to extract the shale oil *in situ*. Two such processes under development are depicted in Figure 101.

Crushed shale is burned from the top down, and oil is collected at the bottom. This oil is extremely viscous, however, and its viscosity must be reduced before it can be easily transported through a pipeline.

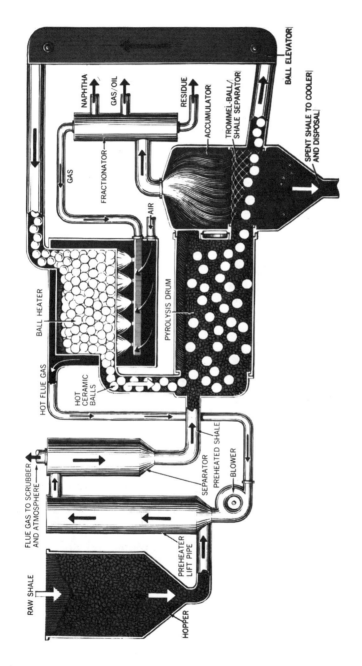

Figure 100. Drawing of TOSCO's (The Oil Shale Co.) retort used to release oil from shale.[2] (Reproduced from *Popular Science* with permission Times Mirror Magazines, Inc. and Ray Pioch, artist.)

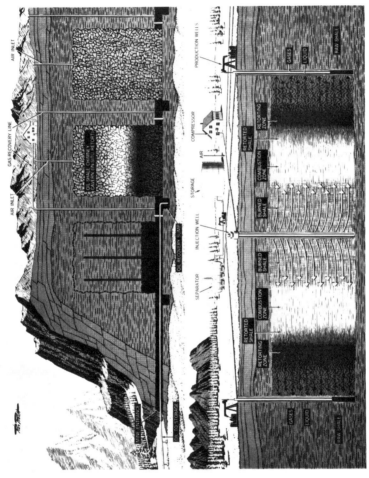

Figure 10I. Artist's sketch of possible *in situ* shale oil recovery operations. (Reproduced from *Popular Science* with permission Times Mirror Magazines, Inc. and Ray Pioch, artist).

HYDROGEN

A general awareness of the potential of hydrogen as a future fuel has been developing over recent years. It is not intended that energy derived from the combustion of hydrogen replace nuclear energy or other nonfossil energy sources. In fact, large amounts of energy would be required to produce hydrogen in large quantities. On the other hand, it does offer the potential of meeting the transportation and storage problems associated with the use of other energy sources; it could play the role of an energy transfer mechanism.

Several studies of a "hydrogen economy" have been conducted in recent years and many papers on the subject have been published. These studies cover all aspects of a switch to hydrogen as a universal fuel, including the economics, technology, and applications. There is general agreement that hydrogen, particularly that produced by electrolysis of water, could solve our long-range fuel problems. A hydrogen economy fuel system is depicted in Figure 102.

Chemical Hydrogen Production

Hydrogen has been produced as an industrial chemical for many decades. Of the 5055 billion SCF produced worldwide in 1968, 95% was used as a chemical intermediate in the production of ammonia, methanol, refined petroleum fuels, and other chemicals. A vast majority of the hydrogen produced today is derived from hydrocarbon or carbonaceous fossil fuels and the fuel product contains substantial amounts of CO and CO_2. Many of the process steps and chemical reactions are similar to those discussed in the previous section on coal gasification. The reactants are gaseous or liquid hydrocarbons or coal, and water. In the most common process, methane, the principal constituent in natural gas, is catalytically cracked to produce hydrogen by the reaction

$$CH_4 + H_2O \rightarrow CO + 3H_2$$

The carbon monoxide is shifted by reaction with more steam to produce more hydrogen by the reaction

$$CO + H_2O \rightarrow CO_2 + H_2$$

and the CO_2 is scrubbed away. A serious limitation imposed on these processes is the generally accepted forecast that petroleum and natural gas resources will be substantially depleted early in the 21st century if they are not drastically conserved. Coal-based hydrogen manufacturing processes certainly could meet our gaseous- and liquid-hydrocarbon fuel and hydrogen needs for a much longer time. Ultimately, other production methods relying on water as the hydrogen source will have to be considered.

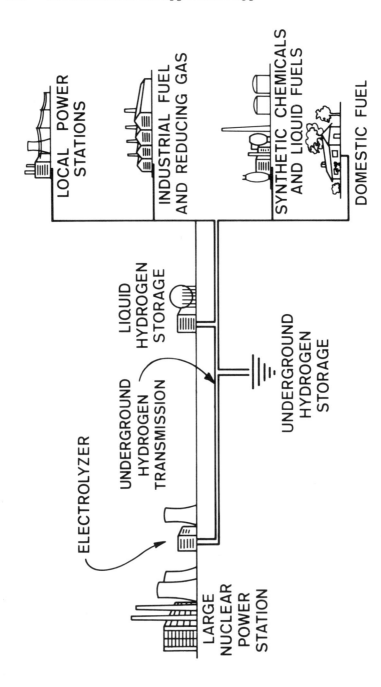

Figure 102. The hydrogen economy fuel system. (Courtesy, D. P. Gregory, Institute of Gas Technology.)

Electrochemical Hydrogen Production

A small fraction of the hydrogen used today is produced by a water-electrolysis process. As electrical energy generated at nuclear power plants becomes more plentiful and fossil fuels become less plentiful, hydrogen production will have to rely increasingly on such a process. Several authors have recently discussed the potential of a combined nuclear-electrochemical hydrogen energy system.[3,4] Recognition is made of the fact that conversion of nuclear energy to electrical energy is most economical when produced in large units (1000 Mw or more), is produced at a constant rate, and that the plants must be located where large cooling requirements can be ecologically met. Due to high costs of transmitting electrical energy over great distances, these authors suggest that this electrical energy be used to produce hydrogen by electrolysis of water.

Electrolysis of water was first accomplished in 1789, but Sir Humphrey Davy has been credited with the first definitive electrolysis in 1806. Commercial production of hydrogen by electrolysis occurs today in locations where inexpensive electric energy is available.

The overall chemical reaction of the electrolysis of water is

$$2H_2O(l) \rightarrow 2H_2\ (g) + O_2(g) \qquad \Delta H = -68{,}300 \text{ cal/gram mole}$$

The process is normally carried out in alkaline solutions due to the corrosiveness of acid solutions. The individual cathode and anode reactions in KOH and NaOH solutions are

$$
\begin{aligned}
\text{Cathode:} \quad & 2H_2O + 2e^- \rightarrow 2H + 2OH^- \\
& 2H \rightarrow H_2(g) \\
\text{Anode:} \quad & 2OH^- \rightarrow 2OH + 2e^- \\
& 2OH \rightarrow H_2O + \tfrac{1}{2}O_2(g) \\
\text{Sum:} \quad & 2H_2O(l) \rightarrow 2H_2(g) + O_2(g)
\end{aligned}
$$

Iron is normally used as a cathode material while anodes are normally made from nickel, cobalt, or nickel-plated steel. Under normal operating conditions, the cell voltage is about 1.25V.

The performance of commercial electrolysis units is affected by the following parameters, cell design, electrode structure, electrolyte concentration, operating pressure, and temperature. Between 125 and 160 kwh of electrical energy per 1000 SCF of hydrogen are required by present-day commercial hydrogen electrolyzers.

Other Methods

Methods other than the chemical and electrochemical methods just discussed also are being studied. On a large scale, production of hydrogen as a primary fuel by conversion of methane does not make sense. Methane is already a superior primary fuel and its decreasing availability precludes such an operation anyway. On the other hand, the cost of producing hydrogen by the electrolysis of water is determined to a large extent by the cost of electricity and the present cost of electrolytic hydrogen is too high, at the present time, to be competitive with other available primary fuels.

The most promising methods for producing hydrogen at lower production costs currently being studied are cyclic thermochemical processes. In such processes the energy to separate hydrogen from water is in the form of heat. In principle, the chemical cycle proceeds as follows:

$$MO_x + H_2O \rightarrow MO_{x+1} + H_2$$

$$MO_{x+1} + (heat) \rightarrow MO_x + \tfrac{1}{2} O_2$$

where M is either a metallic ion or a complex radical. It is thought that the heat could be supplied by a nuclear reactor and the reduced cost of the hydrogen would be based in part on the omission of the intermediate step of converting this heat energy to electrical energy. Many severe development problems remain to be solved in such methods.

Transmission and Storage of Hydrogen

It is anticipated that hydrogen gas can be transmitted in cross-country pipelines in the same manner as natural gas is today. Projections indicate that hydrogen transmission may cost 60% more than natural gas, however. The difference is related to the fact that the heating value of hydrogen is 325 Btu/SCF while that for natural gas is about 1000 Btu/SCF. Thus, to transmit an equivalent amount of energy, about 3 times as much hydrogen as natural gas must be transported. Although the flow capacity of hydrogen nearly compensates for this difference (based on its much lower specific gravity, 0.0695 versus 0.60, and lower viscosity), compression costs for hydrogen will be significantly higher. Estimates for the cost of transporting hydrogen are 1.8–3.8¢/million Btu/100 miles based on early 1970 costs of 1.1–2.4¢/million Btu/100 miles for natural gas.[2] Both of these costs compare favorably to the cost of transmitting electrical energy, 14–19¢/million Btu equivalents/100 miles for 345 KVac systems.[2]

If hydrogen is to serve as an energy transfer mechanism between energy generated at central stations by nuclear means or between the intermittent energy sources (wind, tidal, solar) and various end uses, a storage capability will be necessary. Hydrogen can be stored in either gaseous or liquid form or as a metal hydride. The latter method is currently being studied, and some metal hydrides that contain more hydrogen per unit volume than does liquid hydrogen have been developed. Storage of hydrogen gas in pressurized steel cylinders is common industrial practice. The capacity of storage required in a nuclear-hydrogen economy would probably preclude such storage methods, however. An alternative for storing gaseous hydrogen being considered is underground systems similar to the "gas wells" that contain natural gas. An example of such storage is the 30 billion SCF of helium gas stored at Amarillo, Texas, in a partially depleted natural gas field. Likewise, storage of liquified hydrogen, LH_2, is also common practice. Since the boiling point of hydrogen is very low, $-423°F$, tanks for such storage are generally double-walled with a good vacuum drawn between the walls to reduce conductive heat leakup. The largest LH_2 tank in existence is at the Kennedy Space Flight Center at Cape Canaveral, Florida. It has a capacity of 900,000 gallons and a boil-off of 0.03%/day. In other words, the holding time is greater than nine years. The energy capacity of this tank is about 75% of the capacity of the world's largest pumped hydroelectric storage plant at Ludington, Michigan.

Applications of Hydrogen

It is not the intent here to elaborate on the many end uses of hydrogen. There are numerous publications which illustrate that hydrogen can be used as a general-purpose fuel in nearly all situations where fuel oil, gasoline, or natural gas are used. One application, transportation, was discussed in a previous chapter.

Safety in the handling of hydrogen is the most controversial subject concerning its use. Because of its wide flammability limits and low ignition energy, hydrogen is a hazardous material. This fact would certainly have to be taken into account by designers, and a public education campaign would probably have to be conducted before hydrogen could be widely used. On the other hand, a great deal of experience in handling both gaseous and liquid hydrogen exists and safe handling of these materials has been demonstrated.

METHANE

Methane has already been discussed in conjunction with coal gasification and hydrogen production. However, it is presented separately here because it is one of the principal products of the anaerobic (oxygen-free) digestion of municipal solid wastes.

Anaerobic digestion of organic solid wastes can be carried out in a manner similar to the well established techniques for digesting sewage sludge. The digestion step is one of four different areas of operation in a program of fuel gas production from solid waste. These areas are a) waste handling operations, including separation of nondigestibles and resource recovery, b) digestion, c) gas treatment, including scrubbing and drying, and d) disposal of solid and liquid effluents.

The digestion process requires that the organic matter be mixed with nutrients and various chemicals necessary for proper operation of the digesters. For example, lime and ferrous salts may be added for pH and hydrogen sulfide control, and nutrients added to promote solubilization of the solids. Following this a series of digestion reactions finally yield methane and carbon dioxide.

The amount of organic waste material generated in the United States on an annual basis, which could be converted to methane is truly enormous. As an example, consider the data given in Table 28.

Table 28. Amounts of Dry, Ash-Free Organic Solid Wastes Produced in the United States in 1971.[5]

Source	Wastes Generated (million tons)	Readily Collectable (million tons)
Manure	200	26.0
Urban refuse	129	71.0
Logging and wood manufacturing residues	55	5.0
Agricultural crops and food wastes	390	22.6
Industrial wastes	44	5.2
Municipal sewage solids	12	1.5
Miscellaneous	50	5.0
Total	880	136.3
Net oil potential (10^6 bbl)	1,098	170
Net gas potential (10^{12} ft^3)	8.8	1.36

Chemical analyses of organic solid wastes show that their carbon content is approximately 25% by weight. The net oil potential given in Table 28 is based on conversion of the wastes by reacting carbon monoxide and water giving 1.25 bbl/ton of dry organic wastes. The gas estimate is based on 5.0 cubic feet of methane produced from each pound of organic material. Based on our 1971 consumption of 5.5 billion barrels of oil and 22.8 trillion cubic feet of natural gas, the net potential of these fuels from collectable wastes is significant, 3% and 6% respectively.

ECONOMICS

The technology for producing the "fuels for the future" discussed in this chapter is essentially available. Some development problems remain and many problems associated with scale-up must be solved. Decisions to go ahead with any of these fuels on a major scale depends to a large extent on the economics involved. The economics of large-scale production of synthesized fuels depends on a complex set of factors. Electrical energy costs (particularly for hydrogen), various processing costs, by-product credits, and, of course, raw materials costs are all interdependent.

We will consider one brief example to illustrate the complexity of the problem of projecting costs for synthetic fuels. Natural gas at the wellhead in Texas being supplied under contracts in force prior to 1970 costs $21\text{¢}/10^6$ Btu. By the time this gas was distributed to customers on the east coast the price was $2.00/10^6$ Btu in mid-1975. On the other hand, synthetic natural gas from coal, based on annual operating costs of various synthesizing plants producing a daily total heating value of 250×10^9 Btu, was projected to have a cost at the plant of $1.26/10^6$ Btu in 1973. If transportation and distribution costs are added to this figure, the price of SNG would possibly be 1.5 times that of natural gas. Of course these relative costs are subject to wide fluctuations depending on shifts in the price regulation policies of federal and state governments, and on the rapidly escalating cost of coal, as well as escalating capital and labor costs.

Hydrogen and methane prices are likewise difficult to predict because of the complexity of the interdependence of many factors. Regardless of uncertainties in making economic forecasts at any given point in time, however, we should again remind the reader of the impending depletion of our most desirable fossil fuels—oil and natural gas. Supply and demand of these fuels will certainly play a key role in decisions concerning "fuels for the future."

REFERENCES

1. Kermode, R. I. and J. E. Jones. "Gasification and Liquefaction of Kentucky Coal," Annual Report, Project 2A, Office of Research and Engineering Services, College of Engineering, University of Kentucky, January 1975.
2. Gannon, R. "Shale Oil . . . How Soon?" *Popular Science* 80 (September, 1974).
3. Gregory, D. P., D. Y. C. Ng and G. M. Long. "The Hydrogen Economy," In *Electrochemistry of Cleaner Environments,* J. O'M. Bockris, Ed. (New York, NY: Plenum Press, 1972) p. 226.
4. Gregory, D. P. "Hydrogen as the Universal Fuel," *IEEE INTERCON Technical Papers,* Energy Utilization and Control, 6/5, 1973.
5. Anderson, L. L. "Energy Potential from Organic Wastes: A Review of the Quantities and Sources," Bureau of Mines Information Circular 8549 (Washington, DC: U.S. Department of the Interior, 1972).

SUGGESTED FURTHER READING

1. Bartlit, J. R., F. J. Edeskuty and K. D. Williamson, Jr. "Experience in Handling, Transport and Storage of Liquid Hydrogen — The Recyclable Fuel," Paper 729205, *Proc. of the 7th IECEC,* San Diego, California, September 1972, p. 1312.

2. Booth, L. A., J. D. Balcomb and F. J. Edeskuty. "A Combined Nuclear and Hydrogen Economy — A Long Term Solution to the World's Energy Problem," Paper 739100, *Proc. of the 8th IECEC,* Philadelphia, August 1973, p. 396.

3. "Energy Self Sufficiency: An Economic Evaluation," Policy Study Group of the M.I.T. Energy Laboratory, *Technol. Rev.* **76**, No. 6 (May, 1974).

4. Gregory, D. P. and J. Wurm. "Production and Distribution of Hydrogen as a Universal Fuel," Paper 729208, *Proc. of the 7th Intersoc. Energy Conversion Eng. Conf.,* San Diego, California, September 1972, p. 1329.

5. Hausy, W., G. Leeth and C. Mayer. "Eco-Energy," Paper 729206, *Proc. of the 7th IECEC,* San Diego, California, September 1972, p. 1316.

6. Murray, R. G. and R. J. Schoeppel. "Emission and Performance Characteristics of an Air-Breathing Hydrogen-Fueled Internal Combustion Engine," Paper 719009, 1971, *Proc. of the IECEC,* Boston, August, 1971, p. 47.

7. National Center for Resource Recovery. *Municipal Solid Waste Collection* (Lexington, Massachusetts: Lexington Books, 1973).

8. Stewart, W. F. and F. J. Edeskuty. "Alternate Fuels for Transportation, Part 2: Hydrogen for the Automobile," *Mech. Eng.* (June, 1974), p. 22.

9. Winsche, W. E., T. V. Sheehan and K. C. Hoffman. "Hydrogen— A Clean Fuel for Urban Areas," Paper 719006, 1971, *Proc. of the IECEC, Boston,* August 1971, p. 23.

10. Wiswall, R. H. and J. J. Reilly. "Metal Hydrides for Energy Storage," Paper 729210, *Proc. of the 7th IECEC,* San Diego, Calif, September 1972, p. 1342.

PROBLEMS

1. A commercial coal gasification plant is projected to consume 15,000 tons of coal per day. If the ratio of carbonaceous gases CO, CO_2, and CH_4 produced in the gasifier is 10:10:1, determine the number of pounds of each of these gases produced per day. Assume the coal is 100% carbon.

2. According to Equation (94), the methanation process, common to all coal gasification processes, is highly exothermic. Using the figure obtained in problem 1 for the amount of CO produced per day, determine the number of gallons of cooling water needed per day to remove the heat released. Assume the temperature of the cooling water rises 10°F.

3. a) Look up the density of bituminous coal. Assuming that each of 100 syncrude plants processes 15,000 tons of coal per day, determine the total number of cubic feet of coal per day required for such an operation.

   b) At 100 tons/standard railroad car, how many railroad cars does this amount of coal represent?

4. One cubic foot of water, converted completely to H_2 and O_2 by electrolysis, will produce how many cubic feet of each of these gases? (Assume standard conditions, 1 mole of gas occupies 22.4 liters.)

5. a) Referring to the appendices, determine the energy content, in Btus, of the 5055 billion SCF of hydrogen produced in 1968.

   b) Again in terms of energy content, determine the number of barrels of crude oil equivalent to this amount of hydrogen.

6. The carbon content of organic solid wastes is approximately 25% by weight. Show that the net oil and gas potentials of this material are approximately 1.25 bbl/ton and 10,000 ft^3/ton, respectively.

Chapter 12

Thermionics, Thermoelectrics, and Fuel Cells

INTRODUCTION

In this chapter we will review three different methods of energy conversion. Each of these methods—thermionics, thermoelectrics, and fuel cells—has its own special advantages and problems. None of these methods is currently being used for large-scale energy conversion primarily because of a combination of cost, reliability, and efficiency considerations. In each case, however, there is reason to hope that additional research effort could improve these energy conversion methods to the point where their large-scale commercial application would be possible.

THERMIONICS

Thermionics is the study of the emission of charged particles, usually electrons, from hot surfaces. In 1883, Thomas Edison discovered that an electric current can be generated between two electrodes separated by a vacuum if one of these electrodes is hotter than the other. The Edison effect has been the basis of a very large number of devices, *e.g.*, radio and television vacuum tubes, and has been widely studied. It was O. W. Richardson who suggested, in 1903, that the term thermionic emission be used as a general term for the Edison effect. Figure 103 is a schematic diagram of a thermionic energy converter. Heat supplied to the emitter electrode causes electrons to be "boiled off" this electrode. Thus, the emitter electrode is the cathode. Some of these electrons will be captured by the collector electrode, which is thus the anode. Since the anode is at a lower temperature than the cathode, those collected electrons are trapped. If the anode and cathode are connected electrically via a circuit, the resulting electrical current can be used to do work. A thermionic device is, therefore, a form of heat engine in that it converts heat energy into electrical energy. As such, of course, its maximum possible efficiency (work out divided by heating) will be that given by

$$\text{Maximum Efficiency, } \eta = \frac{T_H - T_L}{T_H} \qquad (95)$$

where T_H is the temperature of the hot (emitter) electrode and T_L is the temperature of the cold (collector) electrode.

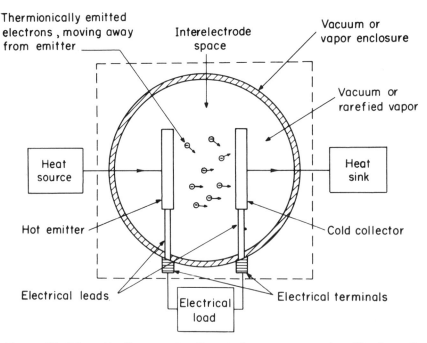

Figure 103. Schematic diagram of a thermionic energy converter. (Courtesy of the M.I.T. Press.)

Clearly, the efficiency of a thermionic converter will also depend on the relative ease with which electrons can be "boiled off" the emitter. The rate of electron emission of electrons from a metal is found to follow an equation of the form

$$I = AT^2 \exp\left(\frac{-eV}{KT}\right) \qquad (96)$$

where $I =$ the rate of electrons being emitted (amp/cm² = coulombs/sec-cm²)

$A =$ The Richardson constant (amp/cm²°K²)
(This constant will be different for every metal)

T = The temperature (°K)

e = The charge on an electron, which amounts to 1.6 x 10^{-19} coulombs/electron

V = the work function (volts) (The value of the work function will also be different for every metal.)

K = Boltzman's constant $(1.38 \times 10^{-23} \frac{joules}{°K})$.

From this equation it is evident that one would like to find a metal for use on the emitter electrode which has a high value of A and a low value of V. Table 29 gives the thermionic properties of some elements and materials. The alkali metal Cs has the lowest work function of any element. This result is not surprising since cesium has the largest measured atom size of any element in the periodic table. To understand why a large atom should give rise to a low work function, consider what the work function represents. The work function (volts) times the charge on an electron (coulombs) gives the work (coulomb-volts or joules) required to remove an electron from the metal. Clearly, this work will depend upon the strength with which the electrons are bound to the metal. In a very simple sense,

Table 29. Thermionic Data for Selected Elements and Materials.

Material	$A(\frac{amp}{cm^2 \ °K^2})$	volts
Barium	60	2.11
Barium strontium oxide	0.5	1.0
Carbon	30	4.34
Cobalt	41	4.41
Copper	—	4.48
Calcium	—	3.20
Cesium	162	1.81
Columbium	37.2	4.01
Hafnium	14.5	3.53
Iron (γ)	1.5	4.21
Iron (a)	26	4.48
Molybdenum	60.2	4.38
Nickel	30	4.61
Osmium	—	4.7
Platinum	17,000	6.27
Rhenium	200	5.1
Rhodium	—	4.58
Tantalum	55	4.19
Tungsten	60.2	4.52
Tungsten + Cs vapor	3	1.5

one might expect this bonding to depend in some way on the atom size. Clearly, from Table 29, this simple view does not explain everything since the lowest work function material in this table is an oxide. This material is BaSrO. Unfortunately, barium strontium oxide has a low value of A, which is itself not surprising since the conductivity of oxides is much lower than that of metals. If a material could be produced with the work function of BaSrO and the Richardson constant of platinum, the whole field of thermionic energy conversion would be drastically altered.

One major problem in the operation of a thermionic converter arises after the electron has left the surface of the emitter. Not all the electrons which have escaped from the emitter are captured by the collector. Those not captured remain between the collector and emitter, and form a space-change barrier that inhibits the emission of additional electrons from the emitter. To counteract this effect, so-called vacuum-diode thermionic converters have been made in which the vacuum gap between the anode and cathode is on the order of a few microns ($25.4\mu = 0.001$ inch). Such extremely close spacing presents many practical difficulties because of the need to heat the emitter to more than $1000°C$ in order to achieve high emitted current densities. At such temperatures the plate which composes the emitter may shift slightly and if this plate makes physical contact with the collector, the device will be shorted and will not function. Langmuir and Kingden discovered in 1923, however, that a tungsten filament immersed in cesium vapor gives a much higher rate of thermionic emission of electrons than would be expected from the values of A and V for pure tungsten. Furthermore, the presence of this cesium vapor also decreased the space-charge barrier effect and allowed the spacing between the emitter and collector to be greatly increased. The explanation for the effect of the presence of cesium vapor involves the fact that cesium, even at high temperatures adsorbs onto the surface of the tungsten. The work function of the resulting, partially Cs covered tungsten surface is found to be a strong function of the degree of coverage of the tungsten surface. This effect is shown in Figure 104. Importantly, as may be seen, the resulting work function can be lower even than the work function of pure cesium.

In addition to its low work function, Cs also has a low ionization potential. The ionization potential is the voltage required to remove an electron from an isolated atom, as opposed to the work function, which is the voltage required to remove an electron from a solid body, not an isolated atom. A hot cesium vapor will, because of its low ionization potential, tend to contain a large number of positive cesium ions. These positive ions in turn, tend to neutralize any space charge effects due to accumulated electrons. The method by which this cesium vapor is produced between the emitter and the collector plates is shown schematically in Figure 105 for the case of a thermionic generator which is heated by radioisotope decay. By adjusting the temperature of the cesium reservoir, the pressure of cesium vapor may be controlled.

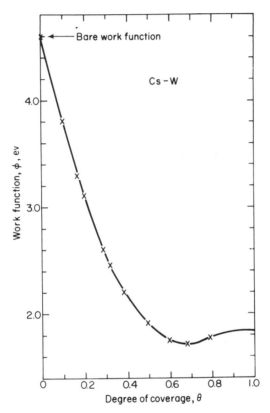

Figure 104. The work function of tungsten as a function of the degree of coverage of the surface with cesium. (Courtesy of the M.I.T. Press.)

Because of the ability of the ionized cesium vapor to neutralize any electron space charge, the emitter and collector plates need not be placed as close together as in the case of the vacuum diode. This increase in spacing combined with the concomitant decrease in work function of the tungsten have made the plasma diode (with a cesiated refractory metal cathode) the most common type of thermionic converter since 1962.

Thermodynamically, a thermionic converter of whatever type is a heat engine. It takes in heat at a high temperature at the emitter and rejects a smaller quantity at the lower temperature of the collector.

The difference in these quantities of energy appears as electrical work. Current thermionic converters have emitter temperatures between 1600 and 2000 °K and have a high power density (5–50 w/cm² of emitter area). Even with collector temperatures on the order of 100s of degrees centigrade, such current thermionic converters could theoretically have efficiencies of up to about 75%. However, current thermionic converters have only moderate efficiencies of between 10–20% due to heat losses at the emitter. Importantly, however, such devices can operate with a variety of heat sources, including nuclear reactors, decaying isotopes, concentrated sunlight, or burners.

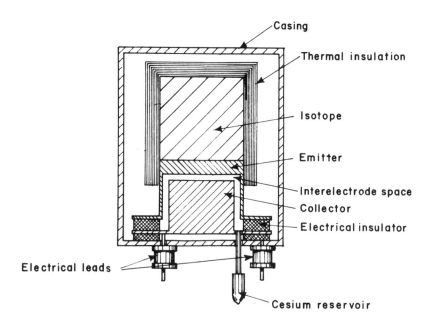

Figure 105. Schematic drawing of a radioisotopic thermionic generator with a cesium source. (Courtesy of the M.I.T. Press.)

Since they do not involve any moving parts, thermionic converters can have very high reliabilities and long lifetimes. Currently, the only major application of such converters, however, has been in the construction of small electrical energy sources that utilize the heat generated by decaying radioactive isotopes to generate electrical power. If materials with higher values of A and V could be discovered, thermionic converters might be used to provide a reliable, efficient way of converting heat to electricity on a large scale. In particular, they could be used in connection with the focusing collectors of solar energy discussed earlier, to convert solar energy to electrical energy. As we will see in the next section, an entirely different type of heat engine could also be used in this way.

THERMOELECTRIC ENERGY CONVERSION

Thermoelectricity is another phenomenon by which thermal energy is converted directly to electrical energy. It depends upon a phenomenon known as the Seebeck effect. Stated simply, the Seebeck effect describes the fact that dissimilar materials joined in a circuit at both a hot and cold junction will generate a voltage. When this device is suitably connected to an external electrical load, an electric current will be generated.

Thermoelectric devices are similar to thermionic devices in that both are heat engines. Thus, they can be treated by thermodynamic second law analyses and are Carnot efficiency limited. Present thermoelectric converters are relatively inefficient (up to one-sixth Carnot), and costly. However, they have the advantages of high reliability (no moving parts) and simplicity. Applications include electric generators for remote areas (weather stations, buoys, space satellites, lunar instrument packages) and medical research.

There are three important effects in thermoelectricity. They are: a) the Seebeck, b) the Peltier, and c) the Thomson effects. The Seebeck effect describes the fact that a voltage is produced in a loop of two different materials, whose junctions are at two different temperatures. This effect is widely used in thermocouples, a dual metal wire device used for measuring temperature. The Peltier effect is the opposite of the Seebeck; that is, a current flowing through a loop of two different materials will result in one junction becoming hot while the other becomes cold. If current is passed through a semiconductor made of a single material, heat is either generated or absorbed in the material and a temperature gradient is established. The magnitude of these effects depends on the direction of flow of the current. This is known as the Thomson effect.

A large number of materials are suitable for thermoelectric operation. Their suitability is determined largely by the product of a parameter, Z, called the figure of merit, and the absolute temperature, T. The figure of merit for a material is defined as

$$Z = \frac{S^2}{k\rho} \tag{97}$$

where S is the absolute Seebeck coefficient for the material (V/°K, positive or negative), K is the thermal conductivity (w/cm°K), and ρ the electrical resistivity (ohm-cm). The units on Z are °K^{-1} and thus the product ZT is dimensionless. The thermoelectric properties of some metals and semiconductors are given in Table 30.

Table 30. Thermoelectric Properties of Selected Elements and Materials.

Material	Electrical Resistivity (ρ, ohm-cm)	Thermal Conductivity (k, w/cm°K)	Seebeck Coefficient (S, V/°K)	Figure of Merit ($Z=S^2/k\rho$, °K^{-1})
Copper	1.67×10^{-6}	3.98	$+ 2.5 \times 10^{-6}$	9.4×10^{-7}
Nickel	6.84×10^{-6}	0.90	-18×10^{-6}	5.5×10^{-5}
Bismuth	1.06×10^{-4}	0.08	-75×10^{-6}	6.6×10^{-4}
Germanium	0.0010	0.636	2.0×10^{-4}	6.3×10^{-7}
Silicon	0.0020	0.835	2.0×10^{-4}	2.4×10^{-5}
Indium antimonide	0.0005	0.170	2.0×10^{-4}	4.7×10^{-4}
Indium arsenide	0.0003	0.315	2.0×10^{-4}	3.8×10^{-4}
Bismuth telluride	0.0010	0.020	2.2×10^{-4}	2.3×10^{-3}

Let us now consider a thermoelectric loop in both the closed and open configurations. Flow of electric charge in a closed loop is shown in Figure 106. The loop consists of n-type and p-type semiconductor elements joined at both ends by a material which is both a good electrical conductor and a good thermal conductor. These semiconductors are typically silicon or germanium doped with an impurity element which in the case of n-type produces an excess of negative charge carriers (electrons) and in the case of p-type, an excess of positive charge carriers (holes). When heat is added, Q_{in}, to the upper junction its temperature is raised to T_H. The positive and negative charge carriers migrate toward the low temperature junction, T_L. At this junction the electrons and holes combine and energy is rejected as heat, Q_{out}. Electrical current, I, will continue to flow as long as heat is added and removed, *i.e.*, as long as a temperature gradient exists.

If the thermoelectric loop is open-circuited, as by removing a segment ΔX from the p element, current will cease to flow and a voltage, ΔV, called the Seebeck voltage, will develop. The magnitude of the voltage is found to be proportional to the temperature difference between the two junctions. That is,

$$\Delta V = S_{pn}\Delta T \qquad (98)$$

for a finite temperature difference, $\Delta T = T_H - T_L$, where $S_{pn} = (S_p - S_n)$ is the temperature dependent proportionality constant for the loop. It is the difference between the absolute Seebeck coefficients of the respective p-type and n-type elements and the subscript, pn, indicates that the current flows from p to n across the cold junction, as shown. It should be noted that the magnitude of the Seebeck

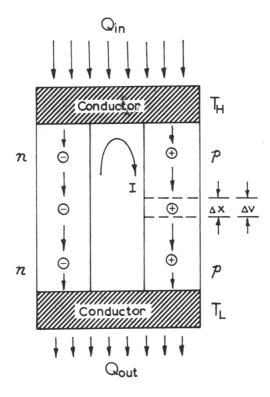

Figure 106. Flow of charge carriers in a thermoelectric loop.

voltage is independent of the geometry of the loop and junctions.

By contrast with the foregoing discussion of Seebeck voltage, if the voltage shown in Figure 106 is supplied by an external source, the flow of charge carriers will be the same and heat required to produce excess change carriers will be absorbed from the surroundings at junction T_H and rejected at junction T_L. This is the principle of a thermoelectric refrigerator. The rate of heat flow at either junction is proportional to the current and is given by

$$Q = \pi_{pn}I \qquad (99)$$

where $\pi_{pn} = (\pi_p - \pi_n)$ is the Peltier coefficient of the thermoelectric loop. The units of π are volts when heat flow is expressed as watt-sec/sec and current, I, is in amperes. The Seebeck and Peltier coefficients are related by $S = \pi/T$.

A thermoelectric loop with the current flowing through a resistive load, R_{Load}, is shown in Figure 107. The voltage drop, ΔV_L, across the external load is equal to the Seebeck voltage, ΔV (Equation 98), minus the voltage losses due to the internal resistance of the p and n semiconductor elements in the loop. The resistance of these elements contain terms which are dependent on both the geometry (length and cross-sectional area) of the element and the inherent resistivities of the materials from which the elements are made. Likewise, the thermal conductance of the loop is dependent on the geometry and the inherent thermal conductives of the elements.

The thermoelectric converter is a heat engine whose power output is given by

$$P = \text{power output} = I^2 R_L \tag{100}$$

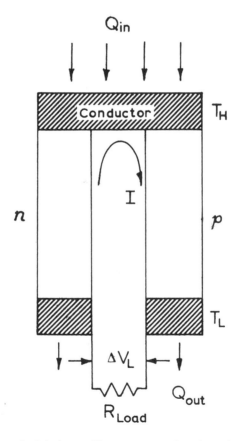

Figure 107. Thermoelectric loop with current passing through an external load.

Keeping in mind that Q_{in} is the energy flow (watt-sec/sec), the thermal efficiency of the converter is given by the relation

$$\eta = \frac{P}{Q_{in}} \tag{101}$$

It can be shown that η is a function of a) ΔT, the temperature difference between the heat source and sink, b) the ratio of the load and internal resistances, $M = R_L/R_C$, and c) the figure of merit, Z, for the converter. The figure of merit for a converter incorporates the geometry of the device, while the figure of merit for a given material is an intensive property, *i.e.*, independent of geometry.

In brief, the thermal efficiency of a specific thermoelectric loop is maximized when M and Z are optimized for a given temperature of operation. The final equation for maximum thermal efficiency is of the form

$$\eta_{max} = fn(Z_{opt}, M_{opt}) \cdot \frac{\Delta T}{T_H} \tag{102}$$

The first term represents a complex function of the parameters Z and M and corresponds to the material efficiency. The second term reflects the fact that a thermoelectric converter is a heat engine and it should be recognized that this term is simply the Carnot efficiency of an engine operating between the temperatures T_H and T_L.

In general, the parameter ZT varies with temperature for both p-type and n-type semiconducting compounds. For any particular material this parameter normally shows a maximum over a fairly narrow temperature range. This is shown schematically in Figure 108. Thus, for given temperature limits, the optimum materials may be readily determined.

Thermoelectric converters are low-voltage, high-current devices. Thus, to obtain large voltages, several converters must be connected electrically in series. Such a configuration is called a thermopile and is depicted in Figure 109. The several converter elements operate within the same temperature range, that is, they are connected thermally in parallel. Thus, electrical current flows through each n and p element successively, while heat flows simultaneously through each element from T_H to T_L.

When a thermoelectric converter is required to operate over a wide temperature range, it can be constructed in either a cascaded or segmented configuration. Both configurations take advantage of the fact that the ZT products for different thermoelectric materials reach a maximum at different temperatures as shown in Figure 108.

A cascaded thermoelectric converter consists of several thermopiles stacked in such a way that the heat rejected by one stage serves

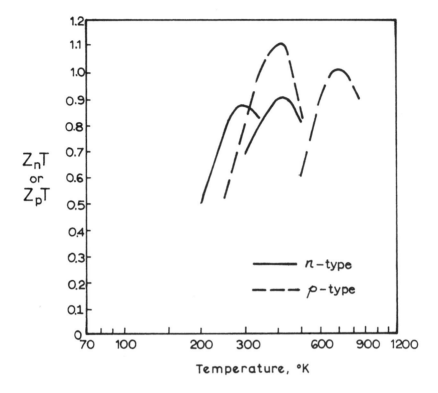

Figure 108. The product ZT versus temperature for representative *n*-type and
    p-type semiconductors.

as the heat source for the stage immediately below it. Adjacent
stages are electrically isolated from one another by a material which
is a good thermal conductor, but is an electrical insulator. Electric
current is drawn from each stage independently. The *p*- and *n*-type
materials in each stage are selected such that their ZT products and
other material properties are optimum for the temperature range in
which the stage will operate.

A segmented converter consists of a single stage. However, each
thermoelectric element is comprised of several different materials
which are, again, selected for optimum performance within the appropriate temperature range.

Among the earliest applications for thermoelectric power were the
SNAP (Systems for Nuclear Auxiliary Power) space power reactors.
A segment of the SNAP-10A thermoelectric converter module is
shown in Figure 110. Heat is brought in from a small nuclear reactor
by hot, molten NaK alloy. This heat is partially converted to electric-

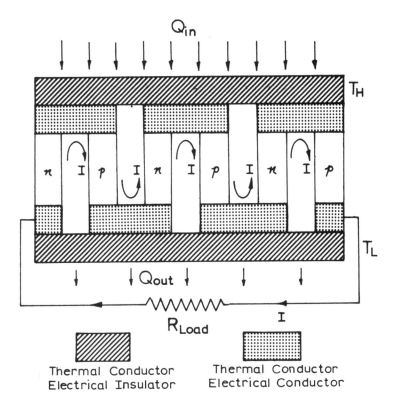

Figure 109. A single-stage thermopile.

ity by the thermoelectric elements shown. Waste heat is radiated to empty space.

SNAP-10A was the first reactor power plant launched into space. Each converter module contained a total of 2880-SiGe thermoelectric elements with a figure of merit of $0.64 \times 10^{-3}{}^{\circ}C^{-1}$. The maximum operating temperature was $1000{}^{\circ}C$. Converter performance compared to design specifications are shown in Table 31.

Table 31. SNAP-10A Performance Characteristics.

Converter Performance	Design	Experimental
Volts/couple, mV	86.5	85.8
Radiator emissivity	0.85	0.89
Element ΔT, ${}^{\circ}F$	280	310
Total converter power, watts	585	560
Weight, lb		146
Life, yr	1	1

Due to materials limitations, it is unlikely that thermoelectric generators with large power outputs will ever be developed. However, thermoelectric generators, using propane fuel, which provide enough power to operate small, portable television sets are currently available.

FUEL CELLS

A fuel cell may be defined as an electrochemical device in which the chemical energy of a conventional fuel is converted directly and usefully into low-voltage direct-current electrical energy. The distinction between fuel cells and primary and secondary batteries should be noted. In the latter, the chemical energy to be converted

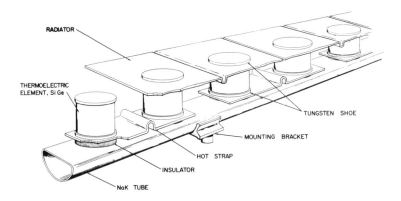

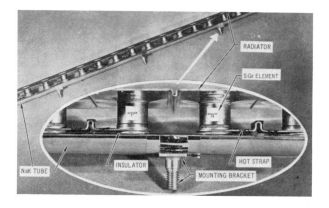

Figure 110. The SNAP-10A thermoelectric converter module. (Courtesy of Atomics International.)

is stored within the cell in the form of metallic anodes, typically lead, zinc, or magnesium. These are certainly not conventional fuels. Fuel cells, on the other hand, are simply energy converters and the fuels—fossil fuels and their derivatives—must be supplied to the cell. Furthermore, in ordinary batteries, the substance reduced at the cathode is typically a compound such as lead oxide, but in fuel cells it is always oxygen, introduced by itself or as a constituent of air.

Fuel Cell Description

A schematic drawing of a hydrogen-oxygen fuel cell is shown in Figure 111. Hydrogen and oxygen (air) are introduced in gaseous form and the following reactions occur:

$$\text{Anode: } H_2(g) + 2OH^- = 2H_2O + 2e^-$$

$$\text{Cathode: } \tfrac{1}{2} O_2(g) + H_2O + 2e^- = 2OH^-$$

$$\text{Sum: } H_2(g) + \tfrac{1}{2} O_2(g) = H_2O(l)$$

The electrons leave the fuel molecules at the anode and do work enroute to being captured by the reaction occurring at the cathode.

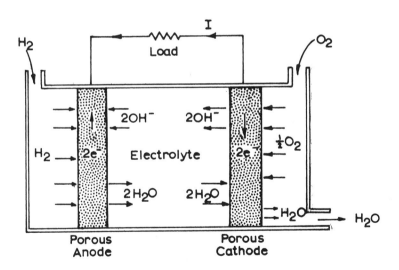

Figure 111. Simple hydrogen-oxygen fuel cell.

There are numerous fuel cell designs and the various components are made from a wide variety of materials. Typically the electrodes are porous and made of metals or carbon, and the electrolyte is an aqueous solution of KOH.

Fuel Cell Performance

The work output of an ideal fuel cell may be expressed in terms of the electron (or ion) transport as

$$W_e = E_r n_e \mathcal{F} = -\Delta G \qquad (103)$$

where E_r is the voltage output of the reversible cell, n_e, the number of electrons transferred per molecule of fuel oxidized, and $\mathcal{F}$ is Faraday's constant:

$\mathcal{F} = 96,500$ coulombs/gram mole $= 96,500$ joules/V-gram mole

$\qquad = 23,060$ cal/V-gram mole.

Faraday's constant is calculated as the product of Avogadro's number N_A and the electronic charge e:

$\mathcal{F} = N_A e = (6.025 \times 10^{23}/\text{gram mole}) (1.602 \times 10^{-19} \text{ coulomb})$
$\qquad = 96,500$ coulomb/gram mole

In terms of operating parameters the free energy change is

$$\Delta G = -E_r I t \qquad (104)$$

where I is the current (amperes) flowing for a time t required to consume one mole of fuel.

The efficiency of an ideal fuel cell is given as

$$\eta_i = \Delta G/\Delta H = -E_r I t/\Delta H \qquad (105)$$

where ΔH is the enthalpy change of the attendant chemical reaction. However, in a real fuel cell the actual voltage E_a is less than that of an ideal cell operating reversibly. Among the reasons for this non-ideality are: a) unwanted anode and cathode reactions may occur, b) normal anode and cathode reactions may be hindered, c) the electrolyte may become inhomogeneous due to concentration gradients, and d) I^2R heating in the electrolyte.

Thus, the actual efficiency will be

$$n_a = -E_a I t/\Delta H$$

The difference between the maximum useful work in reversible operation and the actual work output appears as rejected heat. Even in the absence of irreversibilities, heat is released. Based on the combined first and second laws of thermodynamics, given in Chapter 3, this heat, Q, is determined by the expression

$$Q = T\Delta S = \Delta H - \Delta G \qquad (106)$$

The thermodynamics of a hydrogen-oxygen fuel cell can best be illustrated by an example. Consider such a cell operating at 25°C

(298°K). The standard free energy of formation, $\Delta G°_{298}$, of $H_2O(l)$ is $-56,690$ cal/mole and two electrons are transferred per molecule of H_2. Thus, the cell voltage is

$$E_r = \frac{-\Delta G°_{298}}{n_e \mathscr{F}} = \frac{-(-56,690)}{(2)\,(23,060)} = 1.229 \text{ volts}$$

Since the standard enthalpy (heat) of formation is $\Delta H°_{298} = -68,320$ cal/gram mole, the efficiency of the reversibly operated cell would be

$$\eta_{max} = \frac{\Delta G°_{298}}{\Delta H°_{298}} = \frac{-56,690}{-68,320}\,(100) = 83\%$$

The electrical work output of the cell is

$$W_e = -\Delta G = 56,690 \text{ cal/mole } H_2$$

and the heat which must be removed from the cell is

$$Q = T\Delta S = \Delta H°_{298} - \Delta G°_{298}$$
$$= -68,320 - (-56,690) = -11,630 \text{ cal/mole } H_2$$

In explanation of the high theoretical efficiency of the fuel cell just considered, it should be pointed out that the chemical energy being converted directly to electrical energy is never randomized as heat. Thus, a fuel cell is not a heat engine and the Carnot-cycle limitation imposed on such engines does not apply. The theoretical maximum efficiency is determined solely by the ratio $\Delta G/\Delta H$. Thus, for a fuel cell using pure carbon as a fuel, this would be 100%.

Several requirements must be met in order for a fuel cell to operate at its maximum potential. I^2R losses can be minimized by closely spacing the electrodes and selecting an electrolyte which has high ionic conductivity and negligible electronic conductivity. Other important performance requirements are generally grouped in two categories, reactivity requirements and invariance requirements.

Reactivity requirements relate to the mechanisms and rates of the electrode reactions. Such factors as maintaining proper stoichimetric conditions for complete combustion of the fuel and maintaining high electrolyte activity are important. High-current densities, I/A (A is electrode area), are required for obtaining a ratio of E_a/E_r approaching unity. Reactivity requirements can be met in several ways. Porous electrodes provide an increased gas electrode-electrolyte interface for increased reactivity. Reactivity is also improved by using catalysts either as the electrode or by incorporating them into the electrode material. Increasing operating pressures and temperatures also improve performance.

Invariance requirements are important because the fuel cell is simply an energy converter and if long life is to be assured, certain conditions must be met. Chief among these conditions are the absence of corrosion or side reactions, an invariant electrolyte, and no changes in the electrocatalytic properties of the electrodes. Of course, the reactivity and invariance requirements are interrelated and the establishment of the various cell parameters is a process of compromise and optimization. For example, higher operating temperatures may improve reactivity characteristics but adversely affect electrode invariance.

Fuel Cell Applications

Most fuel cell developments have been aimed at special applications. The most glamorous application of a hydrogen-oxygen fuel cell has been in the Gemini and Apollo space programs. For other space applications such as the Space Shuttle, fuel cells with power outputs of several kilowatts are currently under development. Such power outputs can be obtained by connecting a large number of single cells in series. Typical power output for a single cell is 0.1–0.2 w/cm² (1 volt x·100 to 200 milliamperes per square centimeter of electrode surface).

Possible applications of fuel cells include central power generation and motive power for transportation. In the former case, hydrogen produced by electrolysis of water using excess off-peak electricity could be used during high-demand periods. The energy conversion efficiency of large fuel cells is largely independent of power output and is typically higher than steam and gas turbines, thus making fuel cells attractive for such applications.

The versatility of fuel cells is exemplified by their high specific energy and modest specific power. These properties were shown in Figure 54, and the potential application in transportation is evident.

The principal factors holding back a more widespread use of fuel cell are materials problems which relate directly to economic considerations. These problems include a requirement of expensive platinum group metals for optimum electrode performance in low-temperature cells, or the lack of stable, invariant performance of less-expensive electrode materials otherwise suitable for high-temperature operation.

SUGGESTED FURTHER READING

1. Hatsopoulos, G. N. and E. P. Gyftopoulos. *Thermionic Energy Conversion*, Vol. 1, *Processes and Devices* (Cambridge, Massachusetts: The M.I.T. Press, 1973).
2. Young, G. J. (Ed.) *Fuel Cells* (New York, NY: Van Nostrand Reinhold Co., 1960).

PROBLEMS

1. A thermionic energy converter is made from cesiated tungsten, whose work function is 1.5 volts and whose Richardson constant is 3 amp/cm^2°K. The operating voltage is 1.2 volts. What power is produced at 1500°K by a device having an area of 10 cm^2. Neglect space charge effects.

2. What is the maximum possible current (in amp/cm^2) that can be emitted by a) tantalum, b) tungsten, and c) tungsten in cesium vapor, all at 2000°K?

3. Suppose the device considered in Problem 1 was made from a material having the Richardson constant of platinum and strontium oxide. At 1500°K, what power would be produced if the operating voltage were still 1.2 volts?

4. The suitability of a material for thermoelectric operation is largely determined by the value of the dimensionless product ZT, where Z is the figure of merit for the material. Write an expression for Z, explain all terms, and show that the product ZT is dimensionless.

5. A material, whose Seebeck coefficient varies with temperature according to $S(T) = 10^{-3} + 2 \times 10^{-6}T - 10^{-8}T^2$, is combined with a second material at a junction. When a current of 1 amp is passed through the junction, 0.2 w of heat are absorbed at 300°K. Determine the Seebeck coefficient of the second material and the Seebeck coefficient of the junction at 300°K.

6. Suppose the Seebeck coefficients of the *n*- and *p*-type materials shown in Figure 106 are, respectively, $8 \times 10^{-4} + 10^{-6}T - 1.5 \times 10^{-8}T^2$ and $10^{-3} + 10^{-6}T - 10^{-8}T^2$. If the element operates between the temperatures 1000°K and 500°K, determine a) the open-circuited voltage, and b) the average Seebeck coefficient of the converter.

7. The overall chemical reaction for a fuel cell using carbon (coal) as a fuel would be
$$C + O_2 \rightarrow CO_2 \qquad \Delta H°_{298} = -94.4 \text{ kcal/gram mole}$$
$$\Delta G°_{298} = -94.6 \text{ kcal/gram mole}$$
Calculate the ideal-cell emf and maximum efficiency for such a cell.

8. At 25°C the ideal emf of an H_2-O_2 fuel cell is 1.23 volts. What flow rates of H_2 and O_2 are required to produce a power output of 1 kw?

9. Look up the necessary thermodynamic properties of methane (CH_4) and show that the maximum efficiency for a methane fuel cell is 93%. Show also that the ideal cell voltage is about 1.15 volts. What flow rate in pounds mass per hour of oxygen and methane would be required to produce a power output of 100 kw?

Chapter 13

Planning an Energy Future

A TIME TO CHOOSE

In Chapter 1, several references were made to the Ford Foundation Energy Policy Project. It was a most timely study, beginning as it did in 1972, and culminating in the publication of its final report, *A Time to Choose — America's Energy Future in 1974*. S. David Freeman headed a 39-member professional and support staff, who called upon the talents of some 27 consultants with a broad range of knowledge and experience in the energy field. An elite group of American leaders in the energy industry, education, government policy and regulation, institutions and private organizations concerned with technological and social studies, and the public service companies served as an Advisory Board. They offered counsel throughout the study and reviewed the many drafts leading to the final report. It is highly recommended reading for anyone who would address the question of what kind of energy future we should plan.

There was a near unanimous agreement of all those who participated in the Energy Policy Project[1] with its major conclusion. Namely

> ". . . that it is desirable, technically feasible, and economical to reduce the rate of energy growth in the years ahead . . . Such a conservation oriented energy policy provides benefits in every major area of concern—avoiding shortages, protecting the environment, avoiding problems with other nations, and keeping real social costs as low as possible."[1]

The study recognizes that it would be a mistake to regard energy conservation as an end in itself. This is underscored in the "Statements of the Advisory Board," published with the final report.[2]

> "In looking to the future, the report gives inadequate attention to the question imposed by the finite nature of fossil

271

fuels and what happens when the readily available supplies of oil and gas are exhausted. The approach of that situation in the short term of two or three decades is not presented as a sufficiently serious problem. We believe it deserves more thorough appraisal in terms of its implications for survival of industrial society, for international comity, and for the environmental and economic effects of transition processes that would be triggered long before the marginal resources are depleted."

A smaller group of the Advisory Board states it more specifically as

"But the country must recognize and bear constantly in mind that while conservation will buy time, even the most austere self-discipline will fail to resolve the very real and ultimate problem of supply. The fundamental issue before us is not the simple one so commonly identified as the increase of energy for consumption by a greedy nation. Rather it is the question of whether this nation will be ready with new, practical, economical sources of energy to replace oil and gas when their availability begins to decline."

The choice then is not between whether we should adopt and practice energy conservation measures or develop new energy supplies. We must; in fact it seems abundantly clear that we shall be forced to do both. Rather, the choice is in the time frame of when to initiate new conservation measures, what new sources of supply to pursue, and how far it is possible to go with both before causing severe problems in our economy or losing the backing of the American public.

There is a more or less direct relationship between energy supply and gross national product as has been illustrated in Figure 4. The Ford Energy Policy Project staff believe that there is a good prospect for considerably modifying this relationship through energy conservation practices that rely on improved efficiency of usage. This is a major premise of their recommendation for a reduced energy growth rate. There were two sharply dissenting problems on this premise by Advisory Board members. One was by D. C. Burnham,[3] Chairman, Westinghouse Electric Corporation, who urges prompt development of coal and nuclear electric power plants to meet an energy requirement in the year 2000 greater even than that projected in the Historical Growth Scenario of Figure 7. The second is by M. S. Jameson, Jr.,[3] former executive vice president of the Independent Petroleum Association of America, who believes there is no alternative to an all-out increase in domestic oil and gas production. He contends that a reduced energy growth rate could only come through government edict, and that this would be distasteful and un-American.

One would certainly not expect unanimity of opinion on anything so complex and interwoven in our economical and social order as energy supply. However, the choice of an energy future needs to be made in light of as much knowledge of the consequences of the choice as we can muster, and with the benefit of broad informed public debate.

EXAMPLE OF A SYSTEMS APPROACH METHODOLOGY

Over the past 30 years, a methodology of quantitative assessment for large-scale multi-variable problems has evolved. It is generally known under the name of *Systems Analysis*. Discussions of these sophisticated techniques for modeling of complex problems in technology, business, and social sciences abound in scientific publications; and their results have been prominently displayed in popular literature. One of the more recent additions to this methodology is *Technology Assessment*, a study of the impacts that occur through use of new or expanded technology. TA is a systematic approach to identify and quantify the consequences of the use of technology.

A simple but illustrative example of systems methodology is provided through the results of a NASA/ASEE Systems Design Summer Faculty Program, published under the title MEGASTAR.[4] The objective was to examine postulated energy futures for the United States. Two well-documented energy scenarios were chosen for systems analysis, namely, the Ford Energy Policy Project—Technical Fix Base, and the Westinghouse Nuclear Electric Economy. To these was added a third alternate scenario developed by the participants in the MEGASTAR project.

The energy consumptions projected to the year 2000 by the three energy scenarios are shown in Figure 112. The Westinghouse scenario is for continued energy growth at the rate of the recent past, about 4%/yr. Energy usage would reach 207 Quads in 2000, which is 167% above the 1975 level. The MEGASTAR scenario projects a gradually diminishing growth rate starting at 4%/yr in 1975, and reaching a zero energy growth rate at the end of the century. The Ford Technical Fix Scenario projects an energy growth rate of 1.8%, or one half the historical 3.5% growth rate of 1950–1970. Both the Ford and Megastar projections end at 120 Quads in 2000, a 60% increase over the 1975 usage.*

In Figure 113 the projected source of energy supply for the three scenarios is illustrated.[4] The first bar graph is the actual supply mix

---

*The projection of the Ford Foundation Technical Fix scenario, as presented in its final report was nearer 1.9%, and to a 124 Quad usage in 2000, as shown in Chapter 1. The MEGASTAR participants did not have the final report available, and were apparently using earlier draft reports or solicited information on the Technical Fix Scenario.

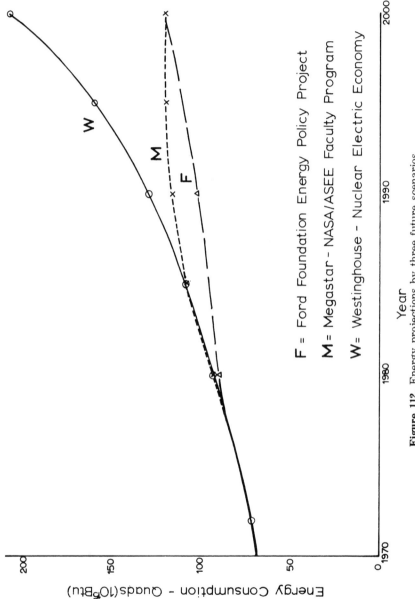

F = Ford Foundation Energy Policy Project

M = Megastar - NASA/ASEE Faculty Program

W = Westinghouse - Nuclear Electric Economy

Figure 112. Energy projections by three future scenarios.

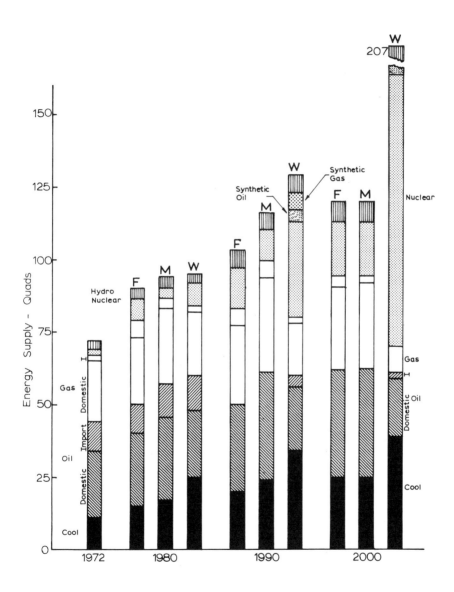

Figure 113. Sources of supply for energy futures: Ford Technical Fix (F), MEGA-STAR (M), and Westinghouse Nuclear Electric Economy (W).

for 1972. Coal, oil, and gas account for 67 of the 72 Quads of energy consumed, or 93%. Oil contributes 33 Quads of which 10 are imported. The projections to 1980 are not much different for the three scenarios; this reflects the fact that it takes considerable lead time, perhaps 7–10 years, to make significant changes in energy supply and energy conversion.

In the years 1990 and 2000 the differences in policy for the scenarios becomes quite evident in Figure 113. The dramatic shift to coal and nuclear projected by the Westinghouse future, is illustrated in the final bar graph at the right of the figure. Out of 207 Quads projected for this Nuclear-Electric Economy, 94 Quads are from nuclear. Another 53 Quads are from coal, both as a direct source for electricity (39) and as the base for synthetic oil and gas (14). Nuclear and coal together account for 71% of the Westinghouse Scenario energy supply in 2000.

A third illustration prepared from the results of the MEGASTAR project[4] is presented as Figure 114. This bar graph is of the projected capital investment required for the three energy scenarios. The capital requirements are broken down into three categories, those for resources (winning the fuel from nature), generation and conversion (primarily for production of electricity), and distribution. For 1973 the actual capital requirements for new and replacement energy facilities was $30 billion, as shown by the small bar at the left of Figure 114.

The capital requirements for the Ford Technical Fix Scenario are modestly above present levels. In the 5-year period 1996–2000, the projected requirement is $185 billion or $37 billion per year. On the other hand, Westinghouse Nuclear Electric Economy requires less capital investment in resources, but a huge total of $414 billion for the period 1996–2000. In that total are $230 billion for energy conversion equipment, and $124 billion for distribution; as shown by the bar graph at the extreme right of Figure 114. The total 25-year requirements for capital investment, steel production, and concrete, as determined by the MEGASTAR system analysis study are given in Table 32.

IMPACTS

To achieve an energy goal of 120 Quads usage in 2000, the 183 Quads projected by the historical growth rate of 3.5%/yr would have to be reduced by 63 Quads. In the Ford Technical Fix Scenario, this reduction would result from energy conservation measures instituted progressively during the 1976–2000 time period. These conservation measures and the attendant savings for the years 1985 and 2000 are given in Table 33.

All of the measures in Table 33 have either no effect or a positive

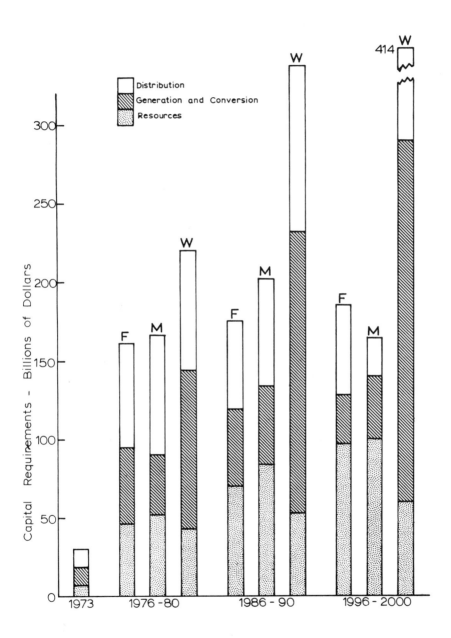

Figure 114. Projected capital requirements for the energy futures of Ford Technical Fix, MEGASTAR, and Westinghouse Nuclear Electric Economy.

Table 32. Capital and Material Requirements,[4] 1976–2000.

Scenario	Capital ($ billions)	Steel (millions tons)	Concrete (millions cu yds)
Ford Technical Fix	867	331	88[a]
MEGASTAR Zero Growth	934	383	342
Westinghouse Nuclear Electric	1621	316	391

[a]Millions of tons of cement.

effect on the environment. Less pollution would result under this scenario. On the other hand there are some economic and social impacts which are extremely negative. Reducing air travel and air freight by switching to rail for short runs, and transferring intercity truck freight to rail are fraught with problems from vested interests and unions. How will railroads, already in financial distress, obtain the needed capital for improved road beds, rolling stock, and terminals? What shall be done about the millions of dollars invested by the trucking industry in big rigs for intercity hauling, and all the union members who drive them?

Other problems arising from the energy savings measures proposed are costs to the homeowners for insulation, heat pumps, and replacement of electric water heaters. Will this be mandated by government or implemented through incentives such as tax breaks or low-cost loans? We cannot offer answers to these questions; they must be studied in the context of other alternatives. Once a decision is made for a policy on energy conservation to be implemented, a gradual transition must be carefully planned. Those economic and social orders that will be disrupted must be given adequate opportunity of preparing for the change. New, satisfactory employment must be found for those who lose jobs by the changes.

The impacts of the Westinghouse Nuclear Electric Economy scenario are very different. To better understand the nature of the impacts consider these paragraphs from the MEGASTAR[6] report.

"Coal use in the NEE scenario increases to six times the present usage by the year 2000, when 40 percent of the coal will be used to make synthetic gas and oil. The total installed electrical capacity will be about 2300 Gw, of which 860 Gw is fossil fired, and 1400 Gw [including 230 Gw in breeders] is nuclear. By the end of the century it is assumed that nuclear capacity will be growing at a constant rate of approximately 100 Gw [or 100,000 Mw] per year, of which one-third will be Liquid Metal Fast Breeders (LMFBR's). Thus, in this sce-

Table 33. Energy Conservation Measures for Technical Fix Scenario.[5]

Measure	Savings over historical growth (Quads)	
	1985	2000
A. **Residential and Commercial Sector**		
1) Space heating - insulation, heat pumps	4	12
2) Air conditioning - insulation, improved efficiency	1	2
3) Water heating - substitute fossil fuel or solar for electric	½	1½
4) Other	½	2
B. **Transportation**		
1) Automobiles - 20 mpg in 1985, 25 mpg in 2000	6	10
2) Aircraft - increase load factor, reduce speeds by 6%, shift short runs to rail service	1½	4
3) Trucks - shift to diesel and intercity traffic to rail	—	2
C. **Industrial**		
1) Energy-intensive industries - improve efficiency	4½	13
2) Process steam and heating - total energy systems and use of heat recuperators	3½	9
3) Other	2½	7½
Totals	24	63

nario, nuclear power grows from 27 Gw in 1974 to 51 times as much in 2000, while fossil fired generators approximately double.

The electric power available to American industries and consumers in the year 2000 will be about five times the present level. Meanwhile the population will have grown from 210 million to 255 million. Some pronounced changes in energy consumption are called for: electric heat pumps for heating and cooling, electric trains, electric automobiles, electrified processes for industry, and more electric machines. Oil and gas consumption in the year 2000, including synthetic fuels, would be down slightly (−7 percent for oil, −20 percent for gas) below current consumption. Gasoline consumption would be down 70 percent."

Such an accomplishment over the next 25 years would call for an all-out national effort. If achieved, we would surely continue to be the world's leader in energy usage per capita, and probably in GNP per capita. Many new advances in technological developments will occur in equipment and methods for nuclear breeder reactors, automated underground mining, strip mining and land restoration, pollution control, electric automobiles, synthetic oil and gas plants, heat pumps, and a variety of new electrical machinery, control, and communication devices. Many engineers would be employed, and several new industries formed.

The negative environmental impacts are numerous. Strip mining would be required on a scale of 5–10 times that of the past. Electric transmission lines would cross many new areas of the countryside, to reach the nuclear generating stations, often located on the coastline or off-shore platforms. In the year 2000 we would be building these stations at a rate equivalent to two 1000-Mw plants in every state of the union per year.

Assuming an optimistic 40% average overall efficiency for operation of the 2300 Gw of generating station capacity installed in the year 2000, and a load factor of 60%, the heat rejected from these stations would be about 62 Quads. The oceans, very large lakes, or massive cooling towers are the only prospects for dissipating this enormous heat release. For example, 14 billion gallons per minute of cooling water, at a 20°F temperature rise, would be required, and this is equivalent to the capacity of Lake Erie every 6.3 days. Thermal pollution of coastal or lake waters seems virtually assured. Wet cooling towers might well release so much water vapor into the air as to have marked changes on local weather.

Almost certainly the most negative aspect of all for the nuclear electric future is the control of fissionable materials. The Ford Foundation Energy Policy Project team devoted much study to the problems of nuclear safety, violence, and waste disposal.[7] They concluded that the element of risk was not well understood, the result of failure terrible in its consequences. The recommendation is for slow growth to buy time in a search for improved methods of control, and for public debate on the nature and consequences of the risks involved.

A major economic impact of the Westinghouse Scenario is in the requirements for capital investment. The debt load to finance a $1620 billion energy program over the next 25 years may be beyond the limits of capability of the country. There will be a requirement for many new workers in the energy field, including an estimated 331,000 engineers, and 2,770,000 nonengineering personnel.[4]

A final impact to consider in choosing such an exponential growth rate as proposed in the Westinghouse scenario, is the global effect of heat release on climate. A summer "Study of Man's Impact on Climate" (SMIC) brought together some of the world experts on the subject at MIT in 1971. The proceedings and a summary are published under the title, *Inadvertant Climate Modification.*[8]

The study identified several feedback mechanisms that may amplify or disperse the effect of additional heat input to the atmosphere. The Stefan-Boltzmann radiation law (*See* Equation (75), Chapter 9) has a negative feedback. More energy release on earth gives an increased temperature, which, in turn, increases radiation to outer space. Suppose world energy usage grows at an average 5%/yr, this would lead to a usage of close to 9000 Quads in the year 2050. This is one-quarter of 1% of the daily solar radiation striking the earth's atmosphere. Direct application of Stefan-Boltzmann leads to a required rise of 0.45°F in the mean global radiating temperature.[9]

The three other feedback mechanisms have generally a positive effect. They are water-vapor and carbon dioxide greenhouse effects, global cloud cover, and polar ice. More energy release on earth leads to a higher average temperature, and greater release of carbon dioxide, water vapor and particulates into the atmosphere. This, in turn, results in more greenhouse effect and melting of polar ice. The albedo or reflective character of the earth to solar radiation is reduced when seawater replaced ice as a surface cover. Thus more solar radiation is absorbed and the increased greenhouse effect holds in the energy at the lower atmospheric level, both contributing to further global temperature rise.

The experts are not in agreement on the magnitude of these feedback effects, but they do agree on the need for caution and more study. SMIC suggests a possible 1°F rise in global temperature by 2000 due to increased CO_2. One participant in that study believes that CO_2 is the contributing cause to the disastrous West African drought.[9] Total melting of Greenland ice would raise the world sea level an estimated 23 ft. Obviously even a fraction of this melting would inundate many coastal areas.

GOVERNMENT POLICY

On one thing nearly everyone is agreed, we need a national energy policy. The question remains what policy? Representative Mike McCormack, Chairman, Subcommittee on Energy Research, Development and Demonstration of the House Committee on Science and Technology; and Chairman, Subcommittee on Environment and Safety of the Joint Committee on Atomic Energy, has enumerated six guiding principles for energy policy.[10]

1) Any energy policy must be based on the best scientific and engineering facts available.

2) A national energy policy must require optimum conservation in everything we do, but especially in the conversion, transmission, and consumption of energy.
3) A national energy policy must allow a large segment of our people to continue to strive for a higher standard of living.
4) A national energy policy must strike a balance. A three-cornered dynamic equilibrium should exist between energy conversion and consumption in one corner, a rational program for protecting the environment and conserving resources in the second, and maintenance of a stable economic system in the third corner.
5) A national energy policy should provide energy self-sufficiency —not by 1980 or 1985, which is totally unrealistic, but as soon as possible and certainly by the year 2000.
6) A national energy policy should provide for an ultimate reliance upon inexhaustible supplies of essentially nonpolluting sources of energy.

A proposal for Project Independence—energy self-sufficiency in the United States by 1980 or 1985—is not considered realistic. In January 1974, shortly after its announcement, David J. Rose[11] advanced telling arguments to show such a policy not only unachievable but probably unwise. He describes a heirarchical assessment of energy which culminates in a "mobile" greatly unbalanced in favor of provision over utilization and conservation. We have modified his "mobile" based on our assessment of the energy picture. Our mobile is shown in Figure 115, and the following statement by Rose applies equally well for our case. Rose writes: "In this arrangement one can see that utilization and conservation on the right half of the mobile, merits as much attention and efforts as all the options dangling in five tiers on the left half of the mobile. Yet one small line at the bottom left, the liquid metal fast breeder reactor represents an option that will receive $320 million in Federal and private development funds this year [1974], 20% of the entire national budget devoted to energy options. What the mobile dramatizes is an embarrassment of blank space, where options are either absent or so poorly formulated that they have received little attention."

Since Rose's article appeared in January, 1974, advances in filling in some of those blank spaces have been made—solar heating and cooling will receive $50 million or more, over a 3-year period. Fusion, geothermal, wind and ocean thermal research programs have also received additional funding.

At this stage, we cannot dismiss alternatives because they have problems, even serious problems. Coal has the undesirable features of mining and pollution problems, but it is our richest national resource—the only fossil fuel we will have in abundance for the next century. We have spent billions of dollars on nuclear technology and

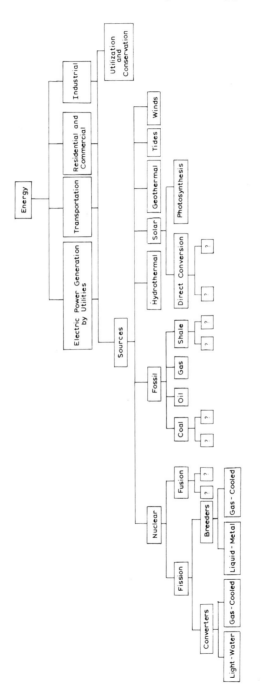

Figure 115. An energy mobile.

are the world leaders in its development for peaceful usage—energy production. It would be foolhardy to abandon or overly restrict its usage without understanding the real nature of the risks in its use and the prospects for reducing those risks to ones we can live with.

Solar energy may sound like the ideal savior for our energy future, and perhaps it will be proved as such in 25 years, or 50, or? But the evidence is not strong enough now to go overboard in that alternative. Costs must be reduced by nearly a factor of 10. Landscapes filled for miles with reflectors and collectors may not be our idea of a very appealing world compared with trees, flowers, grass or even cactus. If we seriously change the albedo of a large area of the earth's surface with solar collectors, it could initiate the same problems of global climate modification caused by polar ice melting.

There are strong vested interests in energy production. Despite their utmost attempts to look at the future in terms of the welfare of our citizens, the company objective is to turn a profit. The easiest way to do this is make more and sell more, preferably of something you already make or modify only slightly. New developments, new explorations, and research, are expensive and risky.

AFTERWORD

In this book we have dealt principally with questions of energy technology, and only in a peripheral way with problems of economics and government policy. We recognize that there are indeed a vast number of political and economic as well as technical problems involved with energy. As time goes on, these problems can be expected to intensify. For example, with regard to present energy resources, economic and government influences can only attempt to divide equitably an every decreasing pie. Looking beyond present resources, however, it seems to us that it is in the area of technology that the best hope for the future lies. It is to the engineering and scientific community that both politicians and economists, as well as everyone else, must look for the long-term solution of our energy problem.

REFERENCES

1. *A Time to Choose - America's Energy Future,* Final Report by the Energy Policy Project of the Ford Foundation (Cambridge, Massachusetts: Ballinger Publishing Co., 1974), p. 325.
2. *A Time to Choose - America's Energy Future,* Final Report by the Energy Policy Project of the Ford Foundation (Cambridge, Massachusetts: Ballinger Publishing Co., 1974), pp. 352–354.

3. *A Time to Choose - America's Energy Future*, Final Report by the Energy Policy Project of the Ford Foundation (Cambridge, Massachusetts: Ballinger Publishing Co., 1974) pp. 362–371; 376–381.

4. *MEGASTAR - The Meaning of Energy Growth: an Assessment of Systems, Technologies and Requirements*, Final Report NASA Grant NGT 01-003-044 (Auburn, Alabama: School of Engineering, Auburn University, 1974).

5. *A Time to Choose - America's Energy Future*, Final Report by the Energy Policy Project of the Ford Foundation (Cambridge, Massachusetts: Ballinger Publishing Co., 1974), pp. 50, 52, 58, 64.

6. *MEGASTAR - The Meaning of Energy Growth: an Assessment of Systems, Technologies and Requirements*, Final Report NASA Grant NGT 01-003-044 (Auburn, Alabama: School of Engineering, Auburn University, 1974), p. 10-1.

7. *A Time to Choose - America's Energy Future*, Final Report by the Energy Policy Project of the Ford Foundation (Cambridge, Massachusetts: Ballinger Publishing Co., 1974), Chapter 8.

8. *Inadvertent Climate Modification - Report of the Study of Man's Impact on Climate (SMIC)*, (Cambridge, Massachusetts: The MIT Press, 1971).

9. Barus, Carl. "Zero Growth: Boon or Bane?", *Mechanical Engineering*, **97**, No. 6, 31 (1975).

10. McCormack, Mike: "A National Energy Policy," *Mechanical Engineering*, **97**, No. 5, 16 (1975).

11. Rose, David J. "Energy Policy in the U.S.," *Scientific American*, **230**, No. 1, 28 (1974).

Appendix 1

CONSTANTS

Acceleration of gravity	g	980.7 m/sec^2
		32.17 ft/sec^2
Atomic mass unit	amu	1.660531 x 10^{-27} kg
Electron rest mass	M_e	9.109558 x 10^{-31} kg
Proton rest mass	M_p	1.672614 x 10^{-27} kg
Neutron rest mass	M_n	1.674920 x 10^{-27} kg
Avogadro's number	N_A	6.022 x 10^{23}/gram-mole
Boltzmann's constant	k	1.38 x 10^{-23} joule/°K
Gas constant	R	1.987 cal/gram-mole °K
		1.987 Btu/lb mole °R
Planck's constant	h	6.62 x 10^{-34} joule•sec
Stefan-Boltzmann constant (black body)	σ	5.67 x 10^{-8} W/m^{2}K^4
Velocity of light	c	2.9979 x 10^8 m/sec
		186,282 mi/sec

PREFIXES

micro (μ) = one millionth
milli (m) = one thousandth
centi (c) = one hundredth
deci (d) = one tenth
deka (D) = ten times

hecta (h) = one hundred times
kilo (k) = one thousand times
mega (M) = one million times
giga (G) = 10^9 times
tera (T) = 10^{12} times

Appendix 2

CONVERSION FACTORS

To convert from	to	Multiply by
Linear Measure		
meters	feet	3.280840
miles (statute)	feet	5280
	kilometers	1.609344
	miles (nautical)	0.868976
Square Measure		
acres	square feet	43560
square feet	square meters	0.092903
square kilometers	square miles	0.386102
square meters	square feet	10.763910
square miles	square kilometers	2.589988
Cubic Measure		
cubic meters	cubic feet	43560
acre-feet	cubic feet	35.314667
American barrels (petroleum)	cubic feet	5.614583
	American gallons	42
	liters	158.98284
American gallons	cubic feet	0.133681
	liters	3.785306
	pints	8
	quarts	4
	American barrels (petroleum)	0.023810
Mass		
kilograms	pounds mass (lbm)	2.204623
pounds	kilograms	0.4535924
metric ton	kilograms	1000
	pounds	2204.6
short ton	kilograms	907.2
	pounds	2000
Time		
calendar year	days	365
	hours	8760
	minutes	5.256×10^5
	seconds	3.154×10^7

To convert from	to	Multiply by
Force		
pound (lbf)	lbm·ft/sec²	1/32.17
	dyne	4.448 x 10⁵
	newton	4.448
newton	kg·m/sec²	1
	dyne	10⁵
	lbf	0.2248
Work and Energy		
Btu	calories (kilogram)	0.252
	foot-pounds	777.649
	horsepower hours	3.92752 x 10⁻⁴
	joules	1054.36
	kilowatt hours	2.92875 x 10⁻⁴
electron volts	joules	1.60209 x 10⁻¹⁹
foot pounds	Btu	1.28408 x 10⁻³
	joules	1.35582
	horsepower hours	5.05050 x 10⁻⁷
	kilowatt hours	3.76616 x 10⁻⁷
joules	Btu	9.48451 x 10⁻⁴
	electron volts	6.24185 x 10¹⁸
	foot-pounds	0.737562
	kilowatt hours	2.7777 x 10⁻⁷
	horsepower hours	3.72506 x 10⁻⁷
kilowatt hours	Btu	3409.52
	horsepower hours	1.34102
	joules	3.6 x 10⁶
Power		
Btu/hour	horsepower	3.92752 x 10⁻⁴
	joules/sec (watts)	1054.35
	kilowatts	2.92875 x 10⁻⁴
horsepower	Btu/hour	2542.48
	foot pounds/sec	550
	joules/sec	745.700
	kilowatts	.745700
kilowatts	Btu/hour	3414.43
	foot pounds/hour	2.65522 x 10⁶
	horsepower	1.34102
	joules/sec	1000
watt	joules/sec	1

Temperature

Degrees Fahrenheit (°F) = 1.8 degrees centigrade (°C) + 32

Degrees Centigrade (°C) = (degrees Fahrenheit (°F) − 32) x 5/9

Absolute Centigrade temperature = °K = °C + 273

Absolute Fahrenheit temperature = °R = °F + 460

Appendix 3

ENERGY RATINGS OF FUELS

FUEL	Btu	Per Unit
Coal:		
Anthracite	12,700	pound
Bituminous	13,100	pound
Coke	12,400	pound
Hydrogen (dry)	325	ft³
	62,050	pound
Natural Gas (dry)	1,035	ft³
	25,047	pound
Natural Gas Liquids (average)	21,325	pound
	4,412,000	barrel
Butane (C₄H₁₀)	21,400	pound
	4,506,000	barrel
Propane (C₃H₈)	21,600	pound
	4,402,000	barrel
Petroleum:		
Crude Oil	5,800,000	barrel
	138,100	gallon
	18,100	pound
Gasoline	5,253,000	barrel
	125,000	gallon
	22,200	pound
Kerosene	5,670,000	barrel
	135,000	gallon
	19,700	pound

FUEL AND ENERGY EQUIVALENTS

1 bbl crude oil	=	443 lb bituminous coal
	=	5,604 ft³ natural gas
	=	1,700 kwh electricity
1 short ton bituminous coal	=	4.52 bbl crude oil
	=	25,300 ft³ natural gas
	=	7,679 kwh electricity
1000 ft³ natural gas	=	79.01 lb bituminous coal
	=	74.95 gallons crude oil
	=	303.34 kwh electricity
1000 kwh electricity	=	1 Mwh electricity
	=	260.5 lb bituminous coal
	=	3,397 ft³ natural gas
	=	0.588 bbl crude oil

INDEX

Index